CAREER COMPOUNDING

A 30-DAYS BLUEPRINT FOR LANDING YOUR DREAM JOB IN THE POWER SECTOR

ANKUR GUPTA

Writing this book has been a labor of love, and I am deeply grateful to everyone who has contributed to its realization.

I extend my heartfelt thanks to my family for their unwavering support and encouragement throughout this journey. Your belief in me has been a constant source of inspiration.

To my mentors and colleagues in the power sector, thank you for sharing your wisdom and expertise. Your insights have enriched the content of this book and shaped my understanding of the industry.

I am also indebted to the countless individuals who graciously shared their experiences and feedback, helping me refine the concepts and strategies presented in these pages.

Special gratitude goes to the team at Notion Press, whose professionalism and dedication have brought this project to fruition.

Last but not least, I extend my sincere appreciation to you, the reader, for choosing to embark on this transformative journey. I hope that this book will serve as a valuable resource on your path to success.

With deepest gratitude,

Ankur Gupta

Contents

Foreword   *vii*

Preface   *ix*

Acknowledgements   *xi*

1. An Introduction - - Igniting Your Career   1

2. Putting Together - Crafting A Power-Packed Resume   5

3. Mastering The Essentials - Basics For Technical Excellence   18

4. Group Dynamics: How To Do Well In Discussions With Others   46

5. Cracking The Interview: How To Do Well In An Interview   60

6. Building A 30-Days : Action Plan   71

7. Week 2: Technical Triumph   85

8. Week 3 : Mastering Group Discussions   109

9. Week 4 - Developing Interview Strategies   112

10. Conclusion: Empowered For Success   126

Key themes and strategies covered in the book include: 133

Foreword

<u>**Final words of encouragement for Celebrating the completion of the 30-days action plan.**</u>

A call to action for readers to share their success stories and join a community of empowered professionals.

Additional resources, templates, and checklists for ongoing success.

- Exclusive access to webinars and workshops hosted by the author.
- Recommendations for further career development and growth in the power sector

Further Reading: Follow me by Scan QR code

Preface

Welcome to "Career Compounding : A 30-Days Blueprint for Landing Your Dream Job in the Power Sector"! As an author deeply passionate about empowering individuals to achieve their career aspirations, it gives me immense pleasure to present this comprehensive guide tailored specifically for those seeking opportunities in the dynamic field of power.

The power sector in India holds vast potential for growth and innovation, offering lucrative career prospects for skilled professionals. However, navigating the job market can be daunting, especially for fresh graduates or those transitioning to new roles. This book is designed to serve as a roadmap, guiding you through every step of the job application process, from crafting a standout resume to acing interviews with confidence.

Drawing upon my own experiences and insights gained from years of working in the industry, I've carefully curated a 30-days action plan that combines practical strategies with actionable tasks. Whether you aspire to join esteemed organizations like NTPC, L&T, BHEL, Adani, or others, this book will equip you with the knowledge and skills needed to succeed.

I encourage you to approach this journey with an open mind and a willingness to invest in your professional development. Remember, success rarely comes overnight, but with dedication, perseverance, and the right guidance, you can achieve remarkable results.

Here's to unlocking your full potential and propelling your career to new heights in the vibrant world of the power sector!

Acknowledgements

The contents of this book, "Career Compounding : A 30-Day Blueprint for Landing Your Dream Job in the Power Sector," are based on the author's personal experiences, insights, and extensive research. The author does not claim to possess proprietary or insider information about any specific organization mentioned in the book.

The thoughts and opinions expressed in this book are solely those of the author and are intended for informational and educational purposes only. The author does not endorse or promote any particular organization, and any references to companies or institutions are made in the context of general knowledge about the power sector.

Readers are encouraged to verify information independently and use their discretion when applying the strategies and advice outlined in this book. The author is not responsible for any actions taken by individuals based on the information provided in this publication.

The author's experiences and anecdotes shared in this book are meant to illustrate broader principles and concepts related to career development in the power sector. They do not imply any direct association with specific organizations or guarantee specific outcomes in the job market.

Disclaimer:

All information provided in this book is derived from publicly available sources, and no attempt has been made to present confidential or proprietary information belonging to any organization. Any resemblance to real persons, living or dead, or actual events is purely coincidental.

This book is designed to serve as a guide for learning and personal empowerment, and readers are urged to tailor the information to their unique circumstances. The author assumes no responsibility for any consequences resulting from the implementation of the strategies outlined in this book.

CHAPTER ONE

AN INTRODUCTION - - IGNITING YOUR CAREER

Two pals, Rahul and Aarav, lived in Delhi long ago. They were engineering students at the same university, and they both had big plans for the future: to work their way up the power sector. They were both excited to start their careers.

After finishing college, Rahul and Aarav went in separate directions. In contrast to Aarav, who looked for advice and focused his efforts, Rahul decided to go it alone in the job market, depending only on his gut and chance.

After some time had gone, they ran across each other again at a class reunion. With his newfound success, Aarav was raking in the dough and advancing his career in the power sector. On the flip side, Rahul was still uncertain about where to stand and what to do next.

Rahul approached Aarav in an effort to uncover the key to his success because he was curious about it. Aarav smiled broadly as he recounted his life's path, detailing how he had followed a set of instructions to achieve his goals.

"How did you manage to achieve so much while I'm still struggling?"

"Aarav," replied Rahul, If you want to know what's important, Rahul, it's having the appropriate direction and making educated decisions, according to Aarav. From a young age on, I learned that achievement is not a result of luck but of persistent effort directed toward a defined objective.

"I stayed focused on my ambitions in the power industry, followed a structured plan, and sought guidance," he continued. I was able to get closer to my goals because I took my time with each step polished with experience of others.

The value of a guided tour became clear to Rahul as he listened carefully. He was moved to action by Aarav's words and decided to seize control of his career.

So, my dear reader, you, too, have the chance to light a fire under your career and plot a path to greatness in the power business, just like Rahul and Aarav did. This book will be your trustworthy companion on a 30-day quest to self-actualization and goal-attainment.

Do you feel prepared to get right in? Join me as we embark on this adventure.

There are good job opportunities in the Indian power sector for people who want to work in a business that is always changing. I am excited to lead you on a 30-day journey that will give you the skills, information, and confidence you need to do well. I have worked on various projects worth over INR 2000 Crores and have a lot of experience as a senior management.

30-Day Action Plan: You're ready to carry out your plan now that you have the skills and information you need. You can stay on track with your job goals if you make a detailed 30-day action plan with daily tasks, exercises, and goals.

- **Understanding the Power Industry**: India's power industry includes a wide range of technologies, such as thermal power plants, renewable energy sources including Solar & Wind, and more. To get around in the industry well, you need to know about the major players, government recruitments portals, and new technical trends.

- **Making Your Career Vision**: It's important to know what you want to do with your career in the power field before you start. Having a clear goal will help you get where you want to go, whether you want to work for well-known companies like NTPC, L&T, or BHEL or help with new renewable energy projects.

- **Resume Reinvention**: The resume you send to possible employers is the first thing they see about you. You can improve your chances of getting interviews and moving up in your job by emphasizing skills, experiences, and accomplishments that are relevant to the power industry.

- **Technical Excellence**: Being good at technology is essential for success in the power industry. It is important for engineers, techs, and analysts to understand basic ideas about how power is generated, transmitted, and distributed.

- **Group Discussion:** Many companies in the power field use group discussions as a way to choose who to hire. To do well in these situations, you need to be able to communicate, work with others, and solve problems.

- **Interview Intensive**: The interview is your chance to show potential companies your skills and experiences. There are insider tips and tricks you can use to ace interviews in the power industry that will help you make a lasting impact.

In the Indian Power sector, including major organizations like NTPC, BHEL, L&T and many more, preferred skills for job aspirants vary based on their experience level and the role they are applying for. Here's a breakdown of preferred skill sets for Entry Level, Middle Level, Experience professional, and Senior Management roles:

1. **Entry Level: (Work Experience: 0 -2 years)**

- **Technical Proficiency:** A strong foundation in technical skills relevant to the power sector, including knowledge of power generation, distribution, and basic engineering principles.

- **Communication Skills:** Clear and effective communication skills, both written and verbal, to convey ideas and work well in a team.

- **Adaptability:** Ability to quickly learn and adapt to new technologies, processes, and industry practices.

- **Problem-Solving:** Basic problem-solving skills and a proactive approach to challenges.

2. **Middle Level:(Work Experience: 2 -5 years)**

- **Specialized Technical Knowledge:** In-depth understanding and application of specialized technical skills relevant to the specific role (e.g., PLC & Logic Diagrams, Piping Engineering , electrical engineering, Renewable Energy, Civil & Structural and many more).

- **Leadership Skills:** Ability to lead a team, delegate tasks, and drive projects to completion.

- **Analytical Thinking:** Strong analytical and critical thinking skills to solve complex problems and make informed decisions.

- **Project Management:** Proficiency in project management tools and methodologies for efficient project execution skills (Primavera, MS Project, SAP, and many more).

3. Experience Professionals : (Work Experience: 5 -10 years)

- **Industry Expertise:** Extensive knowledge of the power sector, market trends, and regulatory frameworks.

- **Strategic Vision:** Ability to contribute to and execute organizational strategies aligned with industry goals.

- **Networking:** Strong industry networking skills to foster relationships with key stakeholders, clients, and industry professionals.

- **Negotiation Skills:** Ability to negotiate and manage contracts, agreements, and partnerships.

4. Senior Management: (Work Experience: 10 years above)

- **Visionary Leadership:** Strong visionary and strategic leadership to guide the organization towards future growth.

- **Decision-Making:** Sound decision-making skills based on a deep understanding of business, market dynamics, and risk assessment.

- **Change Management:** Proficiency in leading organizational change initiatives and adapting to evolving industry landscapes.

- **Global Perspective:** A broad perspective on global energy trends, technologies, and sustainable practices.

It's essential for job aspirants to tailor their resumes and skill sets according to the specific requirements of the roles they are applying for within the Indian Power sector. Additionally, continuous learning and staying updated on industry advancements are crucial at all career levels.

Congratulations on finishing the 30 days' action plan based on role targeted by you, that led you to a rewarding job readiness in the power field. With the skills, information, and confidence you've gained from this guide, you can make a real difference in one of India's most important industries. Best of luck on your journey, and remember to stay focused, resilient, and committed to excellence.

CHAPTER TWO

PUTTING TOGETHER - CRAFTING A POWER-PACKED RESUME

Once upon a time, in a bustling town, there were two best friends named Ram and Shyam. Ram was known for his intelligence and academic excellence. He topped all his subjects and had a detailed knowledge of various software and other engineering skills. Shyam, on the other hand, was more easy-going and friendly, but not as academically inclined as Ram.

One sunny afternoon, as Ram and Shyam were sitting under the shade of a large tree near the college canteen, they began discussing their future aspirations. Both of them shared a dream of securing a good job and making a successful career for themselves.

"I am confident that I'll get any job I apply for," Ram boasted, his eyes sparkling with self-assurance. "After all, I have the best grades and know everything there is to know about my field."

Shyam listened quietly, nodding along. He didn't have the same level of academic achievements as Ram, but he had a different approach to things.

As days went by, Ram began preparing diligently for campus job interviews and technical rounds. He spent hours revising his notes, practicing mock interviews, and polishing his resume. He was so sure of his abilities that he believed he didn't need to make any extra effort.

Meanwhile, Shyam observed his friend's preparations and decided to take a different route. Instead of relying solely on his academic prowess, he decided to focus on tailoring his job applications to each specific role he applied for.

Whenever Shyam found a job listing on the campus notice board that interested him, he carefully read through the requirements and responsibilities mentioned in the job description. He then crafted a resume and cover letter that highlighted his relevant skills and experiences, aligning them with the job requirements and specific to company business goals.

"What's the point of doing all this extra work?" Ram asked one day, noticing Shyam's meticulous approach to job applications. "A good resume is a good resume, no matter how you send it."

Shyam smiled knowingly but didn't say much. He believed in letting his actions speak for themselves.

As weeks passed by, both Ram and Shyam began applying for various job openings in their desired field. Ram sent out his standard resume to every company he applied to, confident that his stellar academic record would speak for itself.

On the other hand, Shyam continued his practice of tailoring his job applications to each specific role. He made sure to address the hiring manager by name in his cover letters and highlighted how his skills and experiences were a perfect match for the job.

One day, to Ram's surprise, he received an email from a company he had applied to. It was a rejection letter, stating that they had decided to move forward with other candidates.

Ram couldn't understand what had gone wrong. He had done everything right, or so he thought. Feeling unhappy, he turned to Shyam for advice.

Shyam listened empathetically to Ram's frustrations before gently sharing his own experiences. He explained how he had been following a different approach to job applications, one that focused on customization and personalization.

"I've been tailoring my resumes and cover letters to match each job description," Shyam revealed. "I address the hiring manager by name and make sure to include specific details about why I'm a good fit for the role. It seems to be working for me."

Ram's eyes widened with realization as he listened to Shyam's words. He had never considered the importance of customization in job applications before. But now, it all made sense.

Feeling inspired, Ram decided to adopt Shyam's approach. He spent the next few days revising his resume and cover letter for each job he applied to, making sure to highlight his most relevant skills and experiences.

To his delight, Ram began receiving interview calls from companies he had previously applied to without success. It seemed that his focus on customization was paying off.

As the days turned into weeks, both Ram and Shyam continued their job search journey. While Ram had initially been skeptical of Shyam's approach, he now realized the importance of personalization and attention to detail in job applications.

Eventually, both friends landed interviews at prestigious companies. Thanks to their hard work, dedication, and valuable lessons learned along the way, they were well-prepared to ace their interviews and embark on successful careers in their chosen field.

In the end, Ram and Shyam learned that success wasn't just about having the best grades or the most impressive resume. It was about being adaptable, open-minded, and willing to learn from each experience. And as they took their first steps into the professional world, they did so with a newfound sense of confidence and purpose, ready to seize whatever opportunities came their way.

In the competitive landscape of the job market, your resume serves as the initial gateway to career opportunities. Crafting a power-packed resume is not merely about listing your qualifications; it's about strategically presenting your skills, accomplishments, and experiences to stand out in the eyes of recruiters. This comprehensive guide aims to demystify the art of resume writing, offering an easy-to-follow step-by-step process tailored for aspiring professionals in thermal power plants and EPC (Engineering, Procurement, and Construction) companies.

Why Your Resume Matters in a Tough Job Market?

In today's tough job market, where employers get a lot of resumes, it's important to have one that stands out. Your resume is like a business tool for you, and how well it works can have a big effect on your chances of getting an interview. This part talks about how important it is to have a well-written resume, focusing on how it can help you make a good first impression and get hired in a competitive process.

Tech and AI tools change quickly, and new tools may have come out since then. I can complete this book, however, give you some general advice on what to put on your resume to make it work better with the AI tools that HR uses. As of my knowledge and suggestions are based on what most people do:

Important Things to Put on Your Resume to Make It Work with AI:

1. **Keywords and Phrases:**

Find keywords that are important to the job from the job description and use them naturally in your resume. AI tools often match resumes based on how relevant the keywords are.

1. **Making changes:**

Make sure that your resume is tailored to each job application. Make the material fit the needs of the job and use different versions of industry-specific words.

3. **Clear Formatting:**

Keep the structure neat and easy to read. To make sure that AI parsing tools can correctly extract information, use standard fonts, and stay away from complicated formatting.

4. **Type in standard sections:**

Put your resume together with standard parts like Contact Information, Skills, Work Experience, Education, and Certifications. This makes it easy for AI tools to sort data into groups.

5. **Skills That Matter:**

Add a section on skills that lists both technical and soft skills that are important. AI often looks for skills that are listed in the job description.

6. **Achievements that can be measured:**

Use measurable measures to draw attention to accomplishments. AI tools can find and highlight successes that can be measured.

7. **The right way to spell and use grammar:**

Make sure you use correct language and spelling. AI tools might not like resumes that have a lot of mistakes on them.

8. Compatible with Applicant Tracking Systems (ATS):

Know what the ATS needs from you. Make sure that your resume is ATS-friendly by using standard headings and staying away from images or tables that might not be parsed properly.

9. Academic Background and Credentials:

Make it clear what you've learned and what certifications you have. AI tools might give more weight to applicants who meet certain school requirements.

10. Consistency:

Formatting should be kept uniform, like using the same date formats and bullet points. This makes sure that AI tools handle information correctly.

Popular AI-Powered Resume Writing Tools

Here is a list of 10 popular free AI tools for effective resume writing. Please note that the popularity and availability of these tools may have changed, so I recommend verifying their current status:

1. Resume Worded:

Offers AI-driven resume feedback and suggestions to optimize content and improve ATS compatibility.

2. SkillSyncer:

Analyzes your resume against job descriptions and provides recommendations to match keywords and skills sought by employers.

3. Jobalytics:

Helps optimize resumes by identifying and integrating relevant keywords from job postings using AI-driven analysis.

4. **RezRunner:**

Evaluate resumes for ATS compatibility and provide suggestions to improve formatting and keyword usage.

5. **Cultivated Culture Resume Scanner:**

Scans resume for ATS compatibility and offers personalized feedback on content and formatting.

6. **VMock:**

Provides AI-powered feedback on resume content, structure, and presentation, along with suggestions for improvement.

7. **Resunate:**

Analyzes resumes for ATS compatibility and offers personalized recommendations to enhance visibility to employers.

8. **Skillroads AI Resume Builder:**

Generates personalized resumes using AI technology and offers insights into keyword optimization and formatting.

9. **TopCV's Free Resume Review:**

Offers AI-driven resume reviews with personalized feedback on content, layout, and overall effectiveness.

10. **Resumonk:**

Provides AI-guided resume templates and suggestions for optimizing content based on industry standards and best practices.

Before using any of these tools, it's essential to review their features, user reviews, and privacy policies to ensure they meet your needs and preferences

<u>Now Understanding the Thermal Power Plants and EPC Industry</u>

Before diving into the intricacies of resume crafting, it's essential to understand the specific requirements of the thermal power plants and EPC industry. This includes knowledge of the key technologies, processes, and skills that are highly valued in these sectors. This section provides a brief overview, setting the stage for tailoring your resume to align with industry expectations.

The cover page of your resume is the first thing recruiters see, and it's crucial to make a lasting impression. This section provides insights into how to effectively structure your cover page for maximum impact related to the Industry focus:

1. Crafting an Engaging Headline:

- Develop a compelling headline that succinctly captures your professional identity.

- Tailor the headline to align with your goals in the thermal power plants and EPC industry.

2. Summary of Qualifications:

- Create a summary section that highlights key qualifications and skills.

- Keep it concise while showcasing your unique value proposition.

3. Tailoring for the Industry:

- Customize your resume for the thermal power plants and EPC industry.

- Use industry-specific keywords to ensure compatibility with recruitment databases.

<u>Sample Cover Letter for</u>

1. "Graduate Engineer job in the Thermal Power Sector"

[Your Name]
[Your Address]
[City, State, Zip Code]
[Your Email Address]
[Your Phone Number]
[Date]
[Hiring Manager's Name]
[Company Name]
[Company Address]
[City, State, Zip Code]
Dear [Hiring Manager's Name],

I am writing to express my keen interest in the [specific position] opportunity within your esteemed organization, as advertised on [where you found the job posting]. With a Bachelor's degree in [Your Engineering Discipline] and a passion for the thermal power sector, I am excited about the prospect of contributing to your team and making a meaningful impact in this dynamic industry.

As an ambitious graduate engineer, I am driven by a strong desire to excel in the thermal power sector and leverage my skills and knowledge to address the challenges and opportunities facing the industry. Throughout my academic journey, I have cultivated a solid foundation in engineering principles, with a particular focus on thermal power generation technologies, renewable energy systems, and sustainable practices.

During my studies, I actively participated in projects and internships that allowed me to gain practical experience and insights into the intricacies of thermal power plants. From conducting feasibility studies for renewable energy projects to analyzing energy efficiency measures, I have honed my analytical skills and developed a holistic understanding of the factors influencing power generation and distribution.

One of my proudest accomplishments is [mention a relevant achievement or project that highlights your skills and capabilities]. This experience not only enhanced my technical proficiency but also instilled in me a sense of responsibility and commitment to excellence. I am confident that my hands-on experience, coupled with my academic background, has prepared me to tackle the challenges of the thermal power sector with confidence and enthusiasm.

What excites me most about the opportunity to join your team is the chance to contribute to innovative projects and collaborate with industry professionals who share my passion for advancing sustainable energy solutions. I am eager to apply my skills in [mention any relevant skills or areas of expertise] to support your company's mission and drive positive change in the thermal power sector.

In addition to my technical abilities, I possess strong communication skills, a collaborative mindset, and a proactive approach to problem-solving. I thrive in fast-paced environments and am adaptable to new challenges and opportunities. I am also committed to continuous learning and professional development, recognizing the importance of staying updated on industry trends and best practices.

I am impressed by [Company Name]'s reputation for excellence and innovation in the thermal power sector, and I am excited about the possibility of contributing to your continued success. I am confident that my passion, drive, and determination make me a strong candidate for the [specific position] role, and I am eager to bring my unique perspective and energy to your team.

Thank you for considering my application. I look forward to the opportunity to discuss how my skills and experiences align with the needs of your organization. Please find my resume attached for your review. I am available at your earliest convenience for an interview and can be reached at [Your Phone Number] or [Your Email Address].

Warm regards,

[Your Name]

2. **Another sample cover page : Piping Engineer Ready to Excel in the Thermal Power Sector**

[Your Name]
[Your Address]
[City, State, Zip Code]
[Your Email Address]
[Your Phone Number]
[Date]
[Hiring Manager's Name]
[Company Name]
[Company Address]
[City, State, Zip Code]
Dear [Hiring Manager's Name],

I am writing to express my enthusiasm for the [specific position] opportunity within your esteemed organization, as advertised on [where you found the job posting]. With a strong background in piping engineering and a passion for the thermal power sector, I am eager to contribute my skills and expertise to your team and make a meaningful impact in this vital industry.

As a dedicated piping engineer, I have always been fascinated by the intricate network of pipelines and systems that form the backbone of thermal power plants. Throughout my academic and professional journey, I have pursued opportunities to deepen my understanding of piping design, construction, and maintenance, with a focus on optimizing efficiency, reliability, and safety.

I hold a Bachelor's degree in Mechanical Engineering with a specialization in Piping Design and Engineering from [Your University]. During my studies, I excelled in courses related to fluid mechanics, thermodynamics, and piping systems, laying a solid foundation for my career in the thermal power sector. Additionally, I actively participated in extracurricular activities and projects that allowed me to apply theoretical concepts to real-world challenges, further honing my practical skills and problem-solving abilities.

One of my most notable achievements is [mention a relevant accomplishment or project that highlights your expertise in piping engineering]. This project not only showcased my technical proficiency but also demonstrated my ability to collaborate effectively with multidisciplinary teams and deliver innovative solutions within tight deadlines.

I am particularly drawn to the opportunity to join your organization because of its reputation for excellence and innovation in the thermal power sector. I am impressed by [Company Name]'s commitment to sustainability, safety, and continuous improvement, and I am eager to contribute to your ongoing success by leveraging my expertise in piping engineering.

In addition to my technical skills, I possess strong communication abilities, leadership qualities, and a proactive mindset. I thrive in dynamic environments where I can collaborate with colleagues and stakeholders to overcome challenges and achieve common goals. I am also committed to staying updated on industry trends and best practices, recognizing the importance of lifelong learning in a rapidly evolving field.

I am excited about the possibility of joining your team and contributing to projects that push the boundaries of innovation in the thermal power sector. I am confident that my passion for piping engineering, coupled with my dedication to excellence and teamwork, make me a valuable asset to your organization.

Thank you for considering my application. I am eager to discuss how my skills and experiences align with the needs of your team and how I can contribute to your organization's success. Please find my resume attached for your review. I am available at your earliest convenience for an interview and can be reached at [Your Phone Number] or [Your Email Address].

Warm regards,

[Your Name]

The Magic of Database Filters: Getting Your Resume Downloaded by HR

Recruitment processes often involve the use of Applicant Tracking Systems (ATS) and database filters. Understanding how these systems work can significantly enhance your chances of getting noticed by HR:

1. Utilizing Keywords Effectively:

- Identify and incorporate relevant keywords from the job description into your resume.

- Optimize your skills and experiences to align with industry-specific terms.

2. Formatting for ATS:

- Ensure your resume is ATS-friendly by using standard fonts, headings, and bullet points.

- Avoid graphics or images that might be misinterpreted by the system.

3. Tailoring for Database Compatibility:

- Fine-tune your resume to match the preferences of ATS used by thermal power plants and EPC companies.

- Regularly update your resume to stay in sync with evolving database algorithms.

4. Networking and Referrals:

- Leverage professional networking platforms to connect with individuals in the industry.

- Seek referrals from professionals working in thermal power plants and EPC companies.

Applicant Tracking Systems (ATS) are used by many Indian companies to manage and streamline the hiring process. To ensure that your resume passes through ATS successfully, you should be aware of common formatting mistakes that can hinder your chances. Here are some common ATS formatting mistakes:

1. Complex or Elaborate Formatting:

- Mistake: Using complex formatting, such as tables, images, or graphics, can confuse ATS systems and lead to parsing errors.

- Solution: Stick to a simple, clean format with standard fonts (e.g., Arial, Calibri) and avoid using text boxes or columns.

2. Incompatible File Formats:

- Mistake: Uploading resumes in formats that ATS may not support, such as PDFs or images.

- Solution: Use a plain text (.txt) or a Microsoft Word (.doc or .docx) file, as these formats are widely supported by ATS.

3. Missing Keywords:

- Mistake: Not incorporating relevant keywords from the job description into your resume.

- Solution: Tailor your resume for each application by including key terms related to the job requirements.

4. Inconsistent Font Usage:

- Mistake: Using multiple fonts or font sizes inconsistently throughout the resume.

- Solution: Maintain a consistent font style and size for headings and body text to ensure easy parsing by ATS.

5. Non-Standard Section Headers:

- Mistake: Using creative or unconventional section headers.

- Solution: Stick to standard section headers like "Work Experience," "Education," and "Skills" to help ATS identify and categorize information correctly.

6. Headers and Footers:

- Mistake: Placing important information, such as contact details, in headers or footers.

- Solution: Include crucial details like your name and contact information in the body of the document to ensure ATS captures the information.

7. Overuse of Special Characters:

- Mistake: Excessive use of symbols, special characters, or non-standard bullets can confuse ATS systems.

- Solution: Keep formatting simple and avoid unnecessary symbols or characters.

8. Excessive Use of White Space:

- Mistake: Leave too much white space, especially in the header or at the end of the document.

- Solution: Ensure a balanced use of white space, and try to fill the page without overcrowding it.

9. Lack of Specificity in Job Titles and Dates:

- Mistake: Using vague job titles or omitting specific dates for employment history.

- Solution: State job titles, include accurate employment dates, and use a consistent date format.

10. Ignoring ATS-Compatible Resume Templates:

- Mistake: Using visually appealing templates that may not be compatible with ATS.

- Solution: Choose a simple and clean resume template that is known to be ATS-friendly.

To maximize your chances of passing through ATS successfully, customize your resume for each application, focus on relevant keywords, and maintain a clean, standardized format.

Crafting a power-packed resume is a nuanced process that involves strategic thinking, attention to detail, and a keen understanding of industry requirements. By following the steps outlined in this comprehensive guide, you can position yourself as a standout candidate in the thermal power plants and EPC sector. Remember, your resume is not just a document; it's your branding, and with the right approach, it can open doors to exciting opportunities in this dynamic industry.

CHAPTER THREE

MASTERING THE ESSENTIALS - BASICS FOR TECHNICAL EXCELLENCE

Understanding the key technical concepts relevant to thermal power plants.

1. Energy Conversion: Thermal power plants convert thermal energy (heat) into mechanical energy and then into electrical energy. For example, heat generated by burning coal in a boiler is used to produce steam, which drives a turbine to generate electricity.

2. First Law of Thermodynamics: This law states that energy cannot be created or destroyed, only transformed from one form to another. In a thermal power plant, heat energy from burning fuel is converted into mechanical energy and then into electrical energy.

3. Second Law of Thermodynamics: This law describes the direction of energy transfer and states that in any energy conversion process, some energy is lost as heat. For example, in a steam turbine, not all heat energy from steam is converted into mechanical energy; some is lost to the surroundings as waste heat.

4. Carnot Cycle: The Carnot Cycle is a theoretical thermodynamic cycle that represents the most efficient heat engine possible. Although not achievable in practice, it serves as a benchmark for comparing the efficiency of actual heat engines.

5. Rankine Cycle: The Rankine Cycle is the thermodynamic cycle used in steam power plants. It consists of four processes: heating water to produce steam (boiler), expanding steam through a turbine to produce work, condensing steam back into water (condenser), and pumping water back to the boiler (pump).

6. Boiler: A boiler is a device that produces steam by heating water. In a thermal power plant, boilers burn fossil fuels such as coal, oil, or natural gas to generate high-pressure steam.

7. Furnace: A furnace is an enclosed structure where the combustion of fuel takes place. In thermal power plants, furnaces provide the heat required to generate steam in boilers.

8. Steam Turbine: A steam turbine is a device that converts the energy of steam into mechanical energy by rotating a shaft. The rotation of the shaft can then be used to drive an electrical generator.

9. Condenser: A condenser is a component in a thermal power plant that condenses steam back into water by transferring heat to a cooling medium (usually water or air). This allows the steam to be reused in the boiler.

10. Cooling Tower: A cooling tower is a structure used to dissipate waste heat from the condenser cooling water into the atmosphere through evaporative cooling. This helps maintain the efficiency of the condenser.

11. Feedwater Heater: A feedwater heater is a device used to preheat water before it enters the boiler. Preheating the feedwater improves the overall thermal efficiency of the power plant by reducing the amount of fuel needed to produce steam.

12. Superheater: A superheater is a component in a boiler that heats steam above its saturation temperature to increase its energy content. Superheated steam has higher enthalpy and is used to drive turbines more efficiently.

13. Reheater: A reheater is a component in a thermal power plant that re-heats steam after it has passed through the high-pressure turbine. Reheating the steam improves the cycle efficiency by reducing moisture content and increasing turbine work output.

14. Economizer: An economizer is a heat exchanger that preheats boiler feedwater using waste heat from the flue gas. This reduces fuel consumption by recovering heat that would otherwise be lost.

15. Steam Generator: A steam generator is another term for a boiler, particularly in nuclear power plants, where nuclear reactions produce heat to generate steam.

16. Turbine Efficiency: Turbine efficiency is the ratio of the actual work output of the turbine to the maximum possible work output. It indicates how effectively the turbine converts steam energy into mechanical energy.

17. Boiler Efficiency: Boiler efficiency is the efficiency of converting fuel energy into steam energy in the boiler. It depends on factors such as combustion efficiency, heat transfer efficiency, and losses in the flue gas.

18. Condenser Efficiency: Condenser efficiency is the ratio of heat removed from the condenser to the maximum possible heat removal. It reflects how efficiently the condenser converts steam back into water.

19. Steam Quality: Steam quality is the fraction of steam in a steam-water mixture. High-quality steam contains a higher fraction of steam molecules and is preferred for efficient turbine operation.

20. Turbine Governing: Turbine governing is the process of controlling the speed and output of the turbine based on power demand. It ensures stable operation of the turbine under varying load conditions.

21. Load Factor: Load factor is the ratio of the average load to the maximum load over a specific period. It indicates how effectively a power plant is utilized over time.

22. Heat Rate: Heat rate is the amount of fuel energy required to produce one unit of electricity. It is a measure of the thermal efficiency of a power plant and is typically expressed in British thermal units (BTU) per kilowatt-hour (kWh).

23. Specific Fuel Consumption: Specific fuel consumption is the fuel consumption per unit of power output. It indicates how efficiently fuel is utilized to generate electricity and is typically expressed in units such as kg/kWh or g/kWh.

24. Ash Handling System: An ash handling system is a set of equipment used to handle and dispose of ash and residues produced by combustion in boilers. It includes equipment for ash collection, transport, and disposal.

25. Coal Handling System: A coal handling system is a set of equipment used to handle and store coal before it is fed into the boiler. It includes equipment such as coal crushers, conveyors, and storage bunkers.

26. Air Preheater: An air preheater is a device that preheats combustion air using heat from the flue gases. Preheating the air improves combustion efficiency and reduces fuel consumption.

27. Pulverized Fuel Firing: Pulverized fuel firing is a method of burning coal in which the coal is ground into fine particles before being fed into the furnace. This allows for efficient combustion and better control of

combustion parameters.

28. Fluidized Bed Combustion: Fluidized bed combustion is a combustion process in which a bed of solid particles is fluidized by an upward flow of air. It is used to burn fuels such as coal, biomass, and waste with high efficiency and reduced emissions.

29. Steam Drum: A steam drum is a component of the boiler where steam is separated from water. It acts as a reservoir for steam and helps regulate steam flow to the turbine.

30. Deaerator: A deaerator is a device used to remove dissolved gases (such as oxygen) from boiler feedwater to prevent corrosion and improve boiler efficiency. Deaerated water is then heated and fed into the boiler.

31. Pressure Relief Valve: A pressure relief valve is a safety device used to release excess pressure in the boiler to prevent over-pressurization and potential damage to equipment.

32. Thermal Efficiency: Thermal efficiency is the ratio of useful energy output to the total energy input, expressed as a percentage. It measures how effectively a power plant converts fuel energy into electricity.

33. Cogeneration: Cogeneration, also known as combined heat and power (CHP), is the simultaneous production of electricity and useful heat from the same energy source. It increases overall energy efficiency by utilizing waste heat for heating or industrial processes.

34. Combined Cycle Power Plant: A combined cycle power plant is a type of power plant that combines gas turbine and steam turbine cycles to achieve higher efficiency. It generates electricity from both the gas turbine exhaust and steam turbine exhaust.

35. Absorption Chiller: An absorption chiller is a device that uses heat energy to produce cooling through the absorption of a refrigerant. It is often used in conjunction with cogeneration systems to provide cooling while utilizing waste heat.

36. Chimney: A chimney is a structure used to expel combustion gases from the boiler to the atmosphere. It helps maintain proper draft and prevent the buildup of harmful gases inside the power plant.

37. Ash Disposal System: An ash disposal system is a system for disposing of ash and residues generated from combustion. It includes methods such as landfilling, ash ponds, or recycling for beneficial use.

38. Boiler Water Chemistry: Boiler water chemistry refers to the treatment of boiler water to prevent corrosion, scaling, and fouling. Proper water chemistry is essential for maintaining boiler efficiency and reliability.

39. Safety Interlocks: Safety interlocks are mechanisms designed to ensure safe operation by preventing unsafe conditions. They may include alarms, shutdown systems, and emergency procedures to mitigate risks in the power plant.

40. Instrumentation and Control Systems: Instrumentation and control systems are systems for monitoring and controlling various parameters in the power plant, such as temperature, pressure, flow, and level. They ensure the safe and efficient operation of the plant.

41. Heat Exchanger: A heat exchanger is a device used to transfer heat from one fluid to another without mixing it. It is used in various components of the power plant, such as condensers, economizers, and feedwater heaters.

42. Stoker Firing: Stoker firing is a method of burning coal in which coal is fed into the furnace by mechanical means, such as conveyor belts or rotating drums. It is commonly used in small to medium-sized boilers.

43. Thermal Pollution: Thermal pollution refers to the release of excess heat into water bodies, causing environmental harm. It can disrupt aquatic ecosystems, reduce dissolved oxygen levels, and impact aquatic life.

44. Power Plant Layout: Power plant layout refers to the arrangement of equipment and structures within the power plant site. It considers factors such as space availability, accessibility, safety, and operational efficiency.

45. Load Dispatching: Load dispatching is the process of allocating power generation from different sources to meet demand. It involves optimizing the operation of power plants to ensure a reliable and economical supply of electricity.

46. Isentropic Efficiency: Isentropic efficiency is the efficiency of a process that occurs with no change in entropy. In the context of turbines and compressors, it represents the ideal efficiency without losses.

47. Energy Storage Systems: Energy storage systems are systems for storing excess energy for later use. They help balance supply and demand, stabilize the grid, and integrate renewable energy sources.

48. Heat Recovery Steam Generator (HRSG): A heat recovery steam generator is a device used to recover waste heat from gas turbine exhaust to generate steam. It is commonly used in combined cycle power plants to improve overall efficiency.

49. Carbon Capture and Storage (CCS): Carbon capture and storage is a technology for capturing carbon dioxide emissions from power plants and storing them underground. It helps mitigate greenhouse gas emissions and

reduce the environmental impact of fossil fuel combustion.

50. Renewable Energy Integration: Renewable energy integration refers to the integration of renewable energy sources like solar and wind into the power grid alongside thermal power plants. It requires coordination of generation, transmission, and distribution to ensure grid stability and reliability.

51. Steam Quality Control: In thermal power plants, maintaining proper steam quality is essential for efficient turbine operation. Steam quality refers to the dryness fraction or the percentage of steam in a steam-water mixture. High-quality steam, with a higher fraction of steam molecules, ensures better turbine performance and reduces the risk of erosion or damage to turbine blades.

52. Turbine Blade Design: Turbine blades play a critical role in converting steam energy into mechanical energy. Design considerations include blade shape, material composition, and cooling mechanisms to withstand high temperatures and pressures. Modern turbines often employ advanced materials like superalloys and ceramic coatings to enhance durability and efficiency.

53. Reheat and Regenerative Cycle: Some thermal power plants incorporate reheat and regenerative cycles to improve overall efficiency. Reheat involves reheating steam at intermediate stages of expansion to maintain high temperatures and prevent condensation, while regeneration preheats feedwater using extracted steam from the turbine to reduce fuel consumption and increase cycle efficiency.

54. Water Treatment and Chemical Conditioning: Proper water treatment is vital to prevent scale formation, corrosion, and fouling in boilers and steam systems. Chemical additives, such as oxygen scavengers, pH adjusters, and scale inhibitors, are used to maintain water quality and prolong equipment lifespan.

55. Environmental Controls and Emissions Reduction: Thermal power plants are subject to stringent environmental regulations governing air emissions, including particulate matter, sulfur dioxide, nitrogen oxides, and mercury. Technologies such as electrostatic precipitators, scrubbers, selective catalytic reduction (SCR), and flue gas desulfurization (FGD) systems are employed to minimize emissions and comply with regulatory standards.

56. Heat Recovery Systems: Heat recovery systems capture waste heat from various plant processes for reuse, improving overall energy efficiency.

Waste heat can be recovered from flue gases, steam turbine exhaust, and cooling water systems to preheat feedwater, generate additional power, or provide heating for industrial processes or district heating networks.

57. Advanced Control and Optimization: Modern thermal power plants utilize advanced control systems and optimization algorithms to maximize efficiency, reliability, and flexibility. Predictive analytics, machine learning, and real-time monitoring help optimize plant operations, manage load fluctuations, and anticipate maintenance needs to reduce downtime and operating costs.

58. Combined Heat and Power (CHP) Systems: Combined heat and power systems, also known as cogeneration, simultaneously produce electricity and useful heat from a single fuel source. CHP systems can achieve overall efficiencies exceeding 80% by utilizing waste heat for heating buildings, industrial processes, or district heating networks, in addition to generating electricity.

59. Flexible Operation and Grid Integration: With the increasing penetration of renewable energy sources and evolving grid dynamics, thermal power plants are adapting to operate flexibly to balance supply and demand. Flexible operation strategies involve ramping up or down output rapidly, providing grid stability services, and integrating with energy storage systems to support intermittent renewables and ensure grid reliability.

60. Life Cycle Assessment (LCA) and Sustainability: Thermal power plants undergo life cycle assessments to evaluate their environmental impacts, resource consumption, and sustainability. LCA considers the entire life cycle of a plant, from raw material extraction and construction to operation, maintenance, and decommissioning, to inform decision-making and identify opportunities for environmental improvement and optimization.

Concepts that are crucial for understanding the design, operation, and maintenance of turbines in thermal power plants, ensuring efficient and reliable power generation.

1. Blade Design: Turbine blades are designed to efficiently extract energy from steam or gas flow. Blade design considers factors such as aerodynamics, material strength, and cooling mechanisms.

2. Blade Erosion and Corrosion: Turbine blades are subjected to erosion and corrosion due to high-velocity steam or gas flow and impurities in the working fluid. Proper maintenance and material selection are crucial to

mitigate these issues.

3. Turbine Inlet Temperature: The temperature of steam or gas entering the turbine affects its efficiency and performance. Higher turbine inlet temperatures result in better efficiency but require advanced materials to withstand the thermal stress.

4. Turbine Exhaust Pressure: The pressure at the turbine exhaust determines the available energy for generating work. Lower exhaust pressures result in higher efficiency but may require larger turbines.

5. Turbine Rotor Dynamics: Turbine rotor dynamics involve the study of rotor vibrations, stability, and balance. Proper rotor design and maintenance are essential to prevent excessive vibrations and ensure safe operation.

6. Steam Turbine Classification: Steam turbines are classified based on their working principles and construction. Common classifications include impulse turbines, reaction turbines, and mixed-flow turbines.

7. Turbine Control Systems: Turbine control systems regulate turbine operation to maintain stability, efficiency, and safety. These systems include governing mechanisms, control valves, and instrumentation for monitoring turbine parameters.

8. Steam Path Components: Steam turbines consist of various components along the steam path, including nozzles, blades, diaphragms, and seals. Each component plays a crucial role in converting steam energy into mechanical work.

9. Condensing vs. Non-Condensing Turbines: Turbines can be classified as condensing or non-condensing based on whether steam is condensed after passing through the turbine. Condensing turbines typically operate at lower pressures and higher efficiencies.

10. Turbine Extraction and Admission: Some steam turbines feature extraction points where steam is tapped off for auxiliary processes such as feedwater heating or process heating. Turbine admission refers to the process of admitting steam at different stages of the turbine for optimal performance.

11. Turbine Overhaul and Maintenance: Turbines require periodic overhauls and maintenance to ensure reliable operation. This includes inspection, repair, and replacement of components such as blades, seals, and bearings.

12. Turbine Efficiency Improvement: Various techniques are employed to improve turbine efficiency, including advanced blade design, inlet steam

conditioning, and steam path optimization.

13. Turbine Bypass Systems: Turbine bypass systems allow for partial steam flow bypassing the turbine during low-load conditions to maintain stable operation and prevent turbine overspeed.

14. Turbine Efficiency Testing: Turbine efficiency is evaluated through performance testing, including heat rate tests, efficiency curves, and thermodynamic analysis.

15. Turbine Trip Systems: Turbine trip systems automatically shut down the turbine in case of abnormal conditions such as overspeed, low oil pressure, or high vibration levels to prevent equipment damage and ensure safety.

16. Turbine Start-up Procedures: Turbine start-up procedures involve preheating, warming up, and gradually increasing steam flow to bring the turbine to operating conditions safely.

17. Turbine Degradation and Aging: Turbines experience degradation and aging over time due to factors such as thermal stress, erosion, and fatigue. Predictive maintenance strategies are employed to monitor turbine health and plan maintenance activities.

18. Turbine Performance Monitoring: Continuous monitoring of turbine performance parameters such as efficiency, vibration levels, and temperature gradients helps identify issues early and optimize operation.

19. Turbine Casing Design: Turbine casings provide structural support and contain steam flow within the turbine. Casing design considers factors such as pressure containment, thermal expansion, and accessibility for maintenance.

20. Turbine Rotor Lock: Turbine rotor lock refers to the condition where the rotor becomes immovable due to thermal expansion or mechanical constraints. Proper design and operational procedures are essential to prevent rotor lock situations.

As Boiler engineering experienced engineers often follow certain guidelines and thumb rules in thermal power plants. Here are 20 thumb rules that are commonly considered important in boiler engineering :

1. Rule of Thumb for Boiler Sizing: Estimate the boiler capacity in horsepower (BHP) using 10-15 BHP per million pounds of steam per hour (lb/hr) load.

2. Water Treatment Rule: For every 1000 pounds of steam generated, use one gallon of boiler water treatment per month.

3. Boiler Efficiency Rule: Regularly monitor and maintain boiler efficiency above 80%, aiming for higher values based on the type and age of the boiler.

4. Blowdown Rule: Set the boiler blowdown rate at 1-3% of the steam generation rate to control impurities and prevent scaling.

5. Stack Temperature Rule: Keep stack temperatures as low as possible to maximize boiler efficiency, typically below 250°F.

6. Feedwater Temperature Rule: Maintain feedwater temperatures at least 20°F below the saturation temperature to prevent thermal shock and improve efficiency.

7. Air/Fuel Ratio Rule: Adjust the air/fuel ratio for optimal combustion efficiency, typically between 10:1 and 20:1.

8. Draft Control Rule: Ensure proper draft control to maintain a slightly negative pressure in the furnace for safety and efficiency.

9. Boiler Drum Level Rule: Maintain the boiler drum level between the normal operating range to prevent carryover and ensure proper water circulation.

10. Fuel Storage Rule: Store fuel away from the boiler in a well-ventilated area with proper safety measures.

11. Furnace Pressure Rule: Keep furnace pressure slightly below atmospheric pressure to avoid air ingress.

12. Boiler Water Chemistry Rule: Regularly monitor and control boiler water chemistry to prevent corrosion, scale, and carryover.

13. Boiler Water Alkalinity Rule: Maintain boiler water alkalinity within the recommended range to prevent acidic corrosion.

14. Sootblowing Rule: Schedule sootblowing based on the cleanliness of heat transfer surfaces, typically every 8-12 hours.

15. Tube Metal Temperature Rule: Keep tube metal temperatures within safe limits to prevent overheating and tube failures.

16. Ash Handling Rule: Optimize ash handling systems to minimize downtime and maintain cleanliness.

17. Condensate Return Rule: Maximize condensate return to the boiler to save energy and reduce water treatment costs.

18. Turbine Bypass Rule: Implement a turbine bypass during low-load conditions to maintain stable turbine operation.

19. Emergency Shutdown Rule: Have a clear and well-practiced emergency shutdown procedure for immediate response to critical situations.

20. Boiler Inspection Rule: Regularly inspect boilers based on a comprehensive schedule to identify and address issues before they become critical.

<u>Practical examples, case studies, and resources for effective preparation</u>

Thermal Power Plant Basics:

1. **What is a thermal power plant?**

A thermal power plant is a facility that generates electricity by converting heat energy into electrical energy through the combustion of fossil fuels or nuclear reactions to produce steam, which drives a turbine connected to an electrical generator.

2. Explain the working principle of a thermal power plant.

A thermal power plant operates by burning fuel (such as coal, natural gas, or oil) to generate heat, which is used to produce steam in a boiler. The high-pressure steam drives a turbine, causing it to rotate and generate mechanical energy. This mechanical energy is then converted into electrical energy by an electrical generator.

3. **What are the main components of a thermal power plant?**

The main components of a thermal power plant include boilers, turbines, condensers, generators, cooling towers, fuel handling systems, ash handling systems, and control systems.

4. **Differentiate between a subcritical and supercritical thermal power plant.**

A subcritical thermal power plant operates at pressures and temperatures below the critical point of water, while a supercritical thermal power plant operates at pressures and temperatures above the critical point. Supercritical plants typically achieve higher efficiency due to increased steam parameters.

5. **Discuss the advantages and disadvantages of thermal power plants compared to other types of power plants.**

Advantages: Relatively low initial investment, high power output, operational flexibility, and established technology.
Disadvantages: Environmental impact (e.g., air and water pollution), dependence on fossil fuels, and potential for greenhouse gas emissions.

Boiler Operations:

6. **How does a boiler work in a thermal power plant?**

A boiler generates steam by heating water using heat energy from burning fuel. The high-pressure steam produced is used to drive a turbine connected to an electrical generator.

7. **Describe the types of boilers used in thermal power plants.**

Common types of boilers used in thermal power plants include pulverized coal-fired boilers, fluidized bed boilers, and nuclear reactors (for nuclear power plants).

8. **What is the purpose of a boiler drum in a thermal power plant?**

The boiler drum acts as a storage vessel for steam generated by the boiler. It helps in separating steam from water and ensures a constant supply of steam to the turbine.

9. **Explain the importance of water chemistry in boiler operation.**

Proper water chemistry is essential to prevent corrosion, scale formation, and carryover in boilers. It involves maintaining appropriate pH levels, controlling dissolved oxygen, and minimizing impurities in the feedwater.

10. **How is boiler efficiency calculated, and what factors affect it?**

Boiler efficiency is calculated as the ratio of useful heat output to heat input. Factors affecting boiler efficiency include combustion efficiency, thermal efficiency, excess air, and heat losses through flue gases and unburnt fuel.

Combustion and Fuel Handling:

11. **What types of fuel are commonly used in thermal power plants?**

Common fuels used in thermal power plants include coal, natural gas, oil, biomass, and in some cases, nuclear fuel.

12. **Discuss the process of fuel combustion in a boiler.**

Fuel combustion in a boiler involves the burning of fuel in the presence of oxygen to produce heat. The heat released raises the temperature of water to generate steam, which is then used to drive turbines.

13. **How is pulverized coal prepared before combustion?**

Pulverized coal is ground into fine particles and mixed with air to form a suspension known as a coal-air mixture. This mixture is then injected into the boiler furnace for combustion.

14. **Explain the role of air preheaters in combustion efficiency.**

Air preheaters preheat the combustion air using waste heat from flue gases. Preheating the air increases its temperature before entering the furnace, which improves combustion efficiency and reduces fuel consumption.

15. **What are the challenges associated with handling biomass fuels in thermal power plants?**

Challenges with biomass fuel handling include variability in fuel composition, moisture content, and ash properties, which can affect combustion efficiency and equipment performance. Specialized handling and storage systems may be required.

Turbine Operations:

16. **Describe the working principle of a steam turbine.**

A steam turbine operates on the principle of converting the kinetic energy of high-pressure steam into mechanical energy by expanding the steam through a series of stationary and rotating blades.

17. **Differentiate between impulse and reaction turbines.**

Impulse turbines utilize the kinetic energy of high-velocity jets of steam to rotate the turbine blades, while reaction turbines utilize both the kinetic energy and pressure of steam to generate rotational motion.

18. **What factors affect the efficiency of a steam turbine?**

Factors affecting turbine efficiency include steam inlet conditions (temperature and pressure), turbine design, blade geometry, internal losses, and steam quality.

19. **Explain the concept of blade erosion and how it is mitigated.**

Blade erosion refers to the gradual wear and tear of turbine blades due to the impact of steam and abrasive particles. It is mitigated through proper material selection, surface coatings, and regular inspection and maintenance.

20. **How is turbine efficiency measured, and what are typical values?**

Turbine efficiency is measured as the ratio of actual work output to the theoretical maximum work output. Typical turbine efficiencies range from 80% to 90%, depending on the type and design of the turbine.

Condenser and Cooling Systems:

21. **What is the purpose of a condenser in a thermal power plant?**

The condenser is used to condense the exhaust steam from the turbine back into water, allowing it to be reused in the boiler. This process increases the efficiency of the power plant by maximizing the use of steam.

22. **Discuss the types of condensers used in thermal power plants.**

Common types of condensers include surface condensers and jet condensers. Surface condensers use cooling water flowing over tubes to condense steam, while jet condensers use water jets to directly condense steam.

23. **How does a cooling tower function, and what are its different types?**

A cooling tower is a heat rejection device that transfers heat to the atmosphere through the evaporation of water. Types of cooling towers include natural draft, mechanical draft (counterflow and crossflow), and hybrid cooling towers.

24. **Explain the impact of cooling water temperature on condenser performance.**

Lower cooling water temperatures result in better condenser performance, as they enhance the heat transfer process and improve the efficiency of steam condensation.

25. **What measures are taken to prevent scaling and fouling in condenser tubes?**

Measures to prevent scaling and fouling include chemical treatment of cooling water, periodic cleaning of condenser tubes, and the use of antifouling coatings on tube surfaces.

Generator and Electrical Systems:

26. **How does an electrical generator work in a thermal power plant?**

An electrical generator converts mechanical energy from the turbine into electrical energy through electromagnetic induction. Rotating coils within a magnetic field produce an alternating current (AC), which is then converted to usable power.

27. **Describe the components of a generator protection system.**

Generator protection systems include overcurrent protection, differential protection, distance protection, and loss of excitation protection to safeguard the generator from electrical faults and damage.

28. **What is a transformer and its role in electrical distribution?**

A transformer is an electrical device that transfers electrical energy between two or more circuits through electromagnetic induction. In a thermal power plant, transformers are used to step up voltage for transmission and step down voltage for distribution.

29. **Discuss the synchronization process of multiple generators.**

Synchronization is the process of connecting multiple generators in parallel with the grid while matching their voltage, frequency, and phase angle. Proper synchronization ensures smooth and safe operation during grid connection.

30. **How are electrical losses minimized in a thermal power plant?**

Electrical losses are minimized through efficient design, use of high-quality conductors, optimal transformer sizing, voltage regulation, and regular maintenance of electrical equipment.

Environmental and Safety Considerations:

31. **What are the environmental impacts of thermal power plants, and how are they mitigated?**

Thermal power plants can contribute to air and water pollution, habitat destruction, and greenhouse gas emissions. Mitigation measures include the use of pollution control technologies (such as scrubbers and electrostatic precipitators), ash disposal management, and transitioning to cleaner energy sources.

32. **Explain the emission control technologies used in thermal power plants.**

Emission control technologies include flue gas desulfurization (FGD) for sulfur dioxide (SO2) removal, selective catalytic reduction (SCR) for nitrogen oxides (NOx) reduction, and particulate matter (PM) control systems such as baghouses and fabric filters.

33. **What safety measures are in place to prevent boiler explosions?**

Safety measures to prevent boiler explosions include regular inspection and maintenance, adherence to safety codes and standards, installation of safety relief valves and pressure gauges, and implementation of operating procedures to prevent overpressure conditions.

34. **Discuss the role of Personal Protective Equipment (PPE) in thermal power plant safety.**

PPE such as hard hats, safety goggles, gloves, earplugs, and respiratory protection is essential for protecting workers from hazards such as heat, chemicals, noise, and airborne particles in thermal power plants.

35. **How is wastewater treated before discharge from a thermal power plant?**

Wastewater from thermal power plants is treated through processes such as sedimentation, filtration, chemical treatment, and biological treatment to remove pollutants before discharge into water bodies or recycling for reuse.

Plant Maintenance and Troubleshooting:

36. **Describe the routine maintenance tasks performed in a thermal power plant.**

Routine maintenance tasks include equipment inspections, lubrication, cleaning, calibration of instruments, replacement of worn components, and testing of safety systems to ensure reliable and safe operation.

37. **What are the common causes of boiler tube failures, and how are they addressed?**

Common causes of boiler tube failures include corrosion, erosion, overheating, and stress corrosion cracking. Addressing these issues involves proper water treatment, material selection, regular inspection, and prompt repair or replacement of damaged tubes.

38. **Discuss the importance of vibration analysis in turbine maintenance.**

Vibration analysis is crucial for detecting mechanical issues such as unbalance, misalignment, bearing wear, and resonance in turbines. It helps prevent catastrophic failures and ensures optimal performance and reliability.

39. **How is a boiler outage planned and executed for maintenance?**

A boiler outage is planned, and tasks are scheduled based on a detailed outage plan. It involves shutting down the boiler, draining fluids, performing inspections, cleaning, repairs, and testing before returning the boiler to service.

40. **Explain the troubleshooting process for a sudden loss of power generation.**

Troubleshooting a sudden loss of power generation involves identifying the root cause through systematic analysis of equipment, instrumentation, and control systems. This may include checking for fuel supply issues, boiler or turbine malfunctions, electrical faults, or grid disturbances.

Control and Instrumentation:

41. **What is a Distributed Control System (DCS), and how is it used in thermal power plants?**

A Distributed Control System (DCS) is a centralized control system used to monitor and control various processes and equipment in a thermal power plant. It integrates data from sensors and actuators to automate plant operations and optimize performance.

42. **Discuss the role of sensors and transmitters in monitoring plant parameters.**

Sensors and transmitters measure parameters such as temperature, pressure, flow rate, and level in different parts of the plant. They provide real-time data for monitoring equipment performance, detecting abnormalities, and ensuring safe operation.

43. **How are control loops tuned for optimal performance?**

Control loops are tuned by adjusting parameters such as proportional, integral, and derivative (PID) gains to achieve stable and responsive control of process variables. Tuning involves iterative adjustment and testing to optimize loop performance.

44. **Explain the function of a Programmable Logic Controller (PLC) in a thermal power plant.**

A Programmable Logic Controller (PLC) is a ruggedized industrial computer used for controlling and automating specific tasks or processes in a thermal power plant. PLCs execute logic based on input signals and control outputs to perform functions such as sequencing, logic control, and data acquisition.

45. **What safety interlocks are in place to protect equipment and personnel?**

Safety interlocks are mechanisms that prevent unsafe conditions by automatically shutting down equipment or activating alarms in response to abnormal situations. Examples include high-pressure and temperature trip systems, flame detectors, and emergency shutdown buttons.

Renewable Integration and Future Trends:

46. **How can renewable energy sources be integrated into a thermal power plant?**

Renewable energy sources such as solar and wind can be integrated into thermal power plants through hybrid systems or co-location of generation facilities. This allows for better grid stability, reduced environmental impact, and increased overall efficiency.

47. **Discuss the concept of a hybrid power plant and its benefits.**

A hybrid power plant combines multiple energy sources, such as thermal, solar, wind, or battery storage, to optimize energy production and grid stability. Benefits include increased reliability, flexibility, and reduced reliance on fossil fuels.

48. **What are the challenges and opportunities of integrating energy storage systems?**

Challenges of integrating energy storage systems include cost, technology limitations, and regulatory barriers. However, energy storage offers opportunities for grid stabilization, peak shaving, and improved utilization of renewable energy resources.

49. **Explain the role of digitalization and automation in the future of thermal power plants.**

Digitalization and automation technologies, such as artificial intelligence, machine learning, and advanced analytics, enable predictive maintenance, optimization of plant operations, and enhanced decision-making for improved efficiency and reliability.

50. **How can thermal power plants adapt to evolving environmental regulations and market demands?**

Thermal power plants can adapt to evolving regulations and market demands by investing in cleaner technologies, implementing emission control measures, enhancing energy efficiency, diversifying fuel sources, and exploring opportunities for renewable integration and energy storage.

Other Renewable energy
Solar Energy: Basics and Fundamentals:
1. What is solar energy, and how is it harnessed?
Solar energy is radiant energy from the sun. It's harnessed through solar panels, which convert sunlight into electricity.
2. Explain the photovoltaic effect.
The photovoltaic effect is the generation of electric current when certain materials are exposed to sunlight, producing a voltage across the material.
3. What are the different types of solar panels used in India?
Monocrystalline, polycrystalline, and thin-film solar panels are commonly used in India.
4. Discuss the significance of solar insolation in India.
Solar insolation refers to the amount of sunlight a region receives. India has high solar insolation, making it ideal for solar energy generation.
5. How do solar tracking systems improve solar energy generation?
Solar tracking systems follow the sun's path, maximizing sunlight exposure on solar panels and increasing energy generation.
Solar Technologies:

6. Describe the working principle of a solar PV system.

Solar PV systems convert sunlight into electricity using semiconductor materials in solar cells.

7. What are the key components of a grid-tied solar PV system?

Solar panels, inverters, mounting structures, and grid connection are key components in a grid-tied solar PV system.

8. Explain the difference between monocrystalline and polycrystalline solar panels.

Monocrystalline panels use single-crystal silicon, offering higher efficiency, while polycrystalline panels use multiple crystals and are more cost-effective.

9. Discuss the advantages and disadvantages of thin-film solar cells.

Thin-film solar cells are lightweight and flexible but have lower efficiency compared to crystalline solar cells.

10. How does a concentrated solar power (CSP) plant operate?

CSP plants concentrate sunlight using mirrors or lenses to generate high-temperature heat, which produces steam to drive a turbine for electricity generation.

Solar Policies and Initiatives:

11. What is the Jawaharlal Nehru National Solar Mission (JNNSM)?

JNNSM is a national initiative promoting solar energy adoption in India to achieve 100 GW of solar capacity by 2022.

12. Discuss the key objectives of the Solar Rooftop Subsidy Scheme.

The scheme aims to encourage solar rooftop installations by providing financial incentives and subsidies to residential and commercial users.

13. Explain the concept of net metering in solar energy.

Net metering allows solar users to feed excess electricity back into the grid, offsetting their consumption and earning credits.

14. What are the financial incentives available for solar energy projects in India?

Financial incentives include accelerated depreciation, tax benefits, and subsidies offered by government schemes.

15. How does the Renewable Purchase Obligation (RPO) promote solar energy adoption?

RPO mandates that a certain percentage of electricity consumed must be from renewable sources, encouraging entities to invest in solar energy.

Solar Projects in India:

16. What are some of the largest solar power plants in India?

Kurnool Ultra Mega Solar Park, Bhadla Solar Park, and Rewa Solar Park are among the largest solar projects in India.

17. Discuss the challenges and opportunities of solar energy integration into the Indian grid.

Challenges include intermittency, but opportunities lie in reducing carbon emissions and enhancing energy security.

18. How does solar energy contribute to rural electrification in India?

Solar microgrids and decentralized solar systems play a crucial role in providing electricity to remote rural areas.

19. Explain the role of solar parks in accelerating solar power deployment.

Solar parks provide large-scale infrastructure for solar projects, optimizing land use and streamlining approvals.

20. What are some innovative solar initiatives at the state level in India?

States like Gujarat and Rajasthan have implemented innovative policies, including solar rooftop programs and solar water pump schemes.

Wind Energy:Basics and Fundamentals

21. What is wind energy, and how is it converted into electricity?

Wind energy is harnessed using wind turbines, which convert kinetic energy from wind into electrical power through turbine generators.

22. Explain the concept of wind turbine aerodynamics.

Wind turbine aerodynamics involves studying the interaction between the blades and the wind to maximize energy extraction.

23. What factors influence wind turbine performance?

Wind speed, air density, turbine height, and rotor diameter are key factors influencing wind turbine performance.

24. Discuss the wind power potential in different regions of India.

Coastal regions, hilltops, and high-wind areas have high wind power potential in India.

25. How do wind shear and turbulence affect wind turbine operation?

Wind shear (change in wind speed with altitude) and turbulence impact the efficiency and stress on wind turbine components.

Wind Turbine Technology:

26. Describe the main components of a horizontal-axis wind turbine (HAWT).

The main components include the rotor, nacelle, gearbox, generator, and tower.

27. What are the advantages and disadvantages of vertical-axis wind turbines (VAWTs)?

VAWTs have simpler designs but lower efficiency compared to HAWTs.

28. Explain the pitch control mechanism in wind turbines.

Pitch control adjusts the angle of the blades to optimize power output and protect the turbine in varying wind conditions.

29. Discuss the concept of variable-speed wind turbines.

Variable-speed turbines adjust rotor speed to optimize energy capture in varying wind conditions.

30. How does a gearbox contribute to wind turbine efficiency?

Gearboxes increase generator speed to match optimal turbine rotation speed for electricity generation.

Wind Policies and Regulations:

31. What is the National Wind-Solar Hybrid Policy in India?

The policy encourages the co-location of wind and solar projects to enhance renewable energy output.

32. Explain the concept of feed-in tariffs (FiTs) for wind energy projects.

FiTs guarantee fixed prices for wind energy, providing revenue certainty for project developers.

33. Discuss the role of competitive bidding in wind power procurement.

Competitive bidding ensures cost-effectiveness and transparency in awarding wind power projects.

34. What are the key provisions of the Wind Energy Program by MNRE?

MNRE's Wind Energy Program aims to promote wind power development through policy support and incentives.

35. How does the Indian Wind Turbine Certification Scheme ensure quality standards?

The certification scheme ensures compliance with international standards, promoting reliable and safe wind turbines.

Wind Projects in India:

36. What are some of the largest wind farms in India?

Muppandal Wind Farm, Jaisalmer Wind Park, and Satara Wind Farm are among the largest wind projects in India.

37. Discuss the challenges and solutions for wind energy grid integration.

Grid stability, curtailment, and transmission infrastructure are key challenges addressed through grid modernization and energy storage.

38. How does wind energy contribute to India's energy security?

Wind energy diversifies the energy mix, reduces dependency on fossil fuels, and enhances energy independence.

39. Explain the concept of repowering in the wind energy sector.

Repowering involves replacing older turbines with newer, more efficient models to increase energy output and operational lifespan.

40. What are some notable offshore wind energy projects planned in India?

Projects like the Gulf of Khambhat and Gulf of Mannar offshore wind projects are in the pipeline to harness wind energy from coastal areas.

Renewable Energy Policies and Initiatives: Government Schemes and Programs:

41. Discuss the key objectives of the PM-KUSUM scheme.

PM-KUSUM aims to promote solar energy adoption in agriculture by installing solar pumps and grid-connected solar projects.

42. What is the DDUGJY, and how does it promote rural electrification through renewables?

DDUGJY focuses on rural electrification, including the deployment of off-grid solar systems and mini-grids.

43. Explain the role of the Green Energy Corridors project in renewable energy integration.

The project aims to strengthen transmission infrastructure for seamless integration of renewable energy into the grid.

44. How does the National Wind-Solar Hybrid Policy encourage hybrid renewable energy projects?

The policy provides incentives and regulatory support for the development of integrated wind-solar projects.

45. Discuss the significance of the ISA for global solar energy cooperation.

ISA promotes solar energy deployment through international collaboration, capacity building, and technology transfer.

Regulatory Framework:

46. What are RECs, and how do they promote renewable energy investment?

RECs certify renewable energy generation, facilitating compliance with renewable purchase obligations and incentivizing investment.

47. Explain the RESCO model for solar project development.

RESCO allows consumers to install solar systems without upfront costs, paying for energy consumed through long-term contracts.

48. What are the regulatory requirements for setting up a renewable energy project in India?

Regulatory requirements include obtaining approvals, permits, and clearances from relevant authorities at the central and state levels.

49. Discuss the role of SERCs in renewable energy regulation.

SERCs regulate tariffs, grid connectivity, and other aspects of renewable energy projects at the state level.

50. How does the Electricity Act, 2003, facilitate renewable energy development in India?

The Act provides a legal framework for renewable energy generation, distribution, and promotion through various mechanisms and incentives.

Basics of Biogas Plants:

51. What is biogas, and how is it produced in biogas plants?

Biogas is a renewable energy source produced through the anaerobic digestion of organic materials such as animal manure, agricultural residues, and organic waste in biogas plants.

52. Explain the anaerobic digestion process in a biogas plant.

Anaerobic digestion is a biological process where microorganisms break down organic matter in the absence of oxygen, producing biogas and digestate as byproducts.

53. What are the key components of a biogas plant?

Key components include a digester tank, inlet and outlet pipes, gas storage unit (e.g., gas holder), mixing system, and gas utilization equipment.

54. Discuss the types of feedstock used in biogas production.

Feedstock can include animal manure, crop residues, food waste, sewage sludge, and energy crops such as maize or sorghum.

55. How does the composition of biogas vary based on the feedstock used?

The composition of biogas varies depending on the feedstock, but typically it consists of methane (50-70%), carbon dioxide (30-50%), and trace amounts of other gases such as hydrogen sulfide and water vapor.

Biogas Plant Design and Construction:

56. Describe the different types of biogas plant designs (e.g., fixed-dome, floating-drum, plug-flow).

Biogas plants can be designed as fixed-dome, floating-drum, plug-flow, or continuous stirred-tank reactors (CSTR), each with unique characteristics and applications.

57. Discuss the factors influencing the selection of biogas plant designs for different applications.

Factors include feedstock type, availability, climate, space constraints, investment costs, and desired biogas production capacity.

58. What are the considerations for site selection and land requirements for a biogas plant?

Site selection should consider proximity to feedstock sources, space for plant infrastructure, soil conditions, access to utilities, and environmental regulations.

59. Explain the construction materials used in biogas plant fabrication.

Common materials include concrete, bricks, steel, and high-density polyethylene (HDPE) for gas storage units and piping.

60. How is the size of a biogas plant determined based on feedstock availability and energy demand?

Plant size is determined by the quantity and type of feedstock available, energy needs for heat and power, and expected biogas production rates.

Feedstock Preparation and Input:

61. Discuss the pre-treatment methods for various feedstocks used in biogas production.

Pre-treatment methods may include shredding, grinding, or mixing to optimize feedstock characteristics such as particle size and moisture content.

62. What are the factors to consider when selecting feedstock for a biogas plant?

Factors include feedstock availability, composition, moisture content, nutrient content, and compatibility with the anaerobic digestion process.

63. Explain the importance of feedstock mixing and ratio optimization in biogas production.

Mixing ensures homogeneity and balanced nutrient ratios, optimizing microbial activity and biogas yield.

64. How do temperature and moisture content affect the anaerobic digestion process?

Temperature and moisture content influence microbial activity and digestion rates, with mesophilic (35-40°C) and thermophilic (50-60°C) conditions preferred for different feedstocks.

65. Discuss the benefits and challenges of using different types of feedstock (e.g., agricultural residues, organic waste) in biogas plants.

Agricultural residues offer high energy potential but may require pre-treatment, while organic waste provides readily available feedstock but may contain contaminants.

Biogas Plant Operation and Maintenance:

66. Describe the startup procedure for a biogas plant.

Startup involves inoculating the digester with microbial culture, gradually introducing feedstock, and monitoring gas production and digester performance.

67. What are the key parameters monitored during the operation of a biogas plant?

Parameters include temperature, pH, gas production rate, gas composition, digester mixing, and hydraulic retention time.

68. Discuss the strategies for optimizing biogas production and enhancing methane yield.

Strategies include optimizing feedstock mixtures, maintaining optimal temperature and pH, maximizing digester mixing, and minimizing hydraulic retention time.

69. Explain the importance of pH control in biogas digester operation.

pH control ensures optimal conditions for microbial activity, with a neutral to slightly acidic pH range (6.5-7.5) typically preferred.

70. How is biogas quality monitored, and what are the safety measures implemented in biogas plants?

Biogas quality is monitored for methane content, hydrogen sulfide levels, and other impurities, with safety measures including gas detection systems, ventilation, and gas storage safety protocols.

GROUP DYNAMICS: HOW TO DO WELL IN DISCUSSIONS WITH OTHERS

Why Group Talks are Important in the Hiring Process:

Group talks or discussions are crucial in the hiring process because they help employers assess how well candidates interact, communicate, and collaborate with others. Employers want to see how candidates contribute to group discussions, listen to others' ideas, and work towards common goals. It reflects teamwork and interpersonal skills, which are essential in

10 effective tips for group discussions, focusing on working together and communicating clearly:

1. **Active Listening:** Pay close attention to what others are saying without interrupting. This shows respect and helps in understanding different perspectives.
2. **Speak Clearly and Confidently:** Express your ideas clearly and confidently to ensure others understand your point of view.
3. **Respect Others' Opinions:** Be open-minded and respectful of diverse viewpoints, even if you disagree. Constructive dialogue fosters a positive atmosphere.

4. **Encourage Participation:** Invite quieter group members to share their thoughts and ideas, ensuring everyone has a chance to contribute.

5. **Stay Focused on the Topic:** Keep discussions on track by addressing relevant points and avoiding tangents or distractions.

6. **Prepare Beforehand:** Familiarize yourself with the topic of discussion and gather relevant information to contribute effectively.

7. **Clarify Doubts:** If you're unsure about something, don't hesitate to ask questions or seek clarification. It demonstrates your engagement and commitment to the discussion.

8. **Use Nonverbal Cues:** Pay attention to nonverbal cues such as body language and facial expressions to gauge understanding and engagement.

9. **Build on Others' Ideas:** Acknowledge and build on the ideas shared by other group members, fostering collaboration and creativity.

10. **Summarize and Synthesize:** Summarize key points and synthesize the discussion to draw conclusions or identify next steps, ensuring clarity and alignment among group members.

Case Studies

1. Actively Listen:

Rahul attentively listened to his coworkers' suggestions during a team meeting at a marketing business without interjecting. He paid close attention to what they had to say about target audiences, messaging, and original ideas. Rahul was able to show his respect for his colleagues' views and his capacity to synthesis differing points of view by combining components from each suggestion into a thorough campaign proposal.

2. Speak With Confidence and Clarity:

Priya had to explain the team's market research findings to top management during a project presentation. Priya didn't stammer or hesitate; instead, she talked with assurance and clarity, supporting her ideas with succinct language and pertinent facts. Her assured speech demonstrated her leadership and communication abilities while assisting the audience in understanding the main points and suggestions.

3. Honor the Thoughts of Others:

Arjun and Ananya, two team members, disagreed on the best strategy for resolving a technical problem at a brainstorming session at a software development company. Even though Arjun and Ananya had different viewpoints, they continued to be courteous and understanding. They had a productive conversation, appreciating one another's viewpoints and working together to explore potential solutions. Their courteous discussion of ideas created a good environment and, in the end, produced an original solution that incorporated aspects of both points of view.

4.   Promote Involvement:

During a weekly meeting of the team at a consulting firm, Rajesh, the team leader, saw that certain team members were reluctant to participate in discussions. Rajesh used a "round-robin" structure to promote teamwork and give everyone a chance to voice their opinions on the items on the agenda. He made sure that everyone felt heard and appreciated and actively encouraged the most reserved individuals to participate. In addition to increasing involvement, this inclusive approach inspired new ideas and insights from team members who might not have spoken up as much in the past.

5.   Remain Adherent to the Subject:

The team began talking about unrelated personal experiences during a project planning session for a building project. Anjali, the project manager, skillfully turned the conversation back to the agenda items after realizing the importance of maintaining focus. She urged the group to put off personal conversations until later and reminded them of the important deadlines and goals they needed to meet. The team was able to sustain productivity and make sure that important decisions were made effectively thanks to Anjali's proactive approach.

Useful Activities to Improve Your Ability to Talk in a Group:

1. **Role-playing exercises:** Practice different scenarios where you have to communicate effectively within a group.

2. **Debates:** Engage in structured debates on various topics to hone your argumentation and persuasion skills.
3. **Group projects:** Collaborate with others on projects or assignments to develop teamwork and communication skills.
4. **Mock interviews:** Participate in mock interviews with peers to simulate real-world group discussion scenarios.

Practical Exercises to Enhance Group Discussion Skills:

1. **Brainstorming sessions:** Practice brainstorming ideas with a group to generate creative solutions to problems.
2. **Case study discussions:** Analyze case studies as a group, discussing possible strategies and outcomes.
3. **Feedback sessions:** Provide and receive constructive feedback on group communication skills to identify areas for improvement.
4. **Team-building activities:** Engage in team-building exercises to foster trust, cooperation, and communication among group members.

50 recent common Group Discussion topics

Climate Change and Environmental Sustainability: Discuss the impact of climate change on the planet and the urgent need for sustainable practices to mitigate environmental degradation. Topics may include renewable energy, conservation efforts, and international agreements like the Paris Agreement.

Women Empowerment: Explore issues related to gender equality, women's rights, and strategies to empower women in various spheres such as education, employment, and leadership roles.

Digital India Initiative: Analyze the government's Digital India initiative aimed at transforming India into a digitally empowered society and knowledge economy. Topics may include digital infrastructure, e-governance, and digital literacy.

Healthcare Infrastructure: Assess the state of healthcare infrastructure in India, challenges in healthcare delivery, and policies needed to improve access, affordability, and quality of healthcare services.

Education Reforms: Examine the education system in India, challenges such as access, quality, and equity, and reforms needed to enhance educational outcomes and promote lifelong learning.

Economic Growth vs. Environmental Conservation: Debate the trade-off between economic growth and environmental conservation, exploring sustainable development models that balance economic prosperity with ecological preservation.

Social Justice and Inequality: Discuss social inequalities based on factors such as caste, religion, gender, and economic status, and strategies to promote social justice, inclusivity, and equal opportunities for all.

Cybersecurity and Data Privacy: Delve into the growing importance of cybersecurity and data privacy in the digital age, addressing threats, vulnerabilities, and regulatory frameworks to safeguard digital information.

Urbanization and Smart Cities: Explore the challenges and opportunities of rapid urbanization, sustainable urban development practices, and the concept of smart cities leveraging technology for efficient governance and resource management.

Political Reforms and Electoral Systems: Evaluate the electoral system in India, electoral reforms, and measures to enhance political accountability, transparency, and fair representation in elections.

Globalization and its Impact: Assess the impact of globalization on India's economy, culture, and society, examining opportunities for economic growth, cultural exchange, and challenges such as inequality and cultural homogenization.

National Security and Border Issues: Discuss national security challenges, border disputes, and strategies to strengthen India's security apparatus, including defense modernization, diplomatic initiatives, and border management.

Crisis Management and Disaster Preparedness: Explore disaster management strategies, preparedness for natural disasters, pandemics, and responses to crises, including coordination among stakeholders and building resilient communities.

Ethical Governance and Corruption: Debate ethical issues in governance, transparency, accountability, and measures to combat corruption in public administration, promoting integrity, and ethical leadership

Cryptocurrency Regulation: Discussing the need for regulating cryptocurrencies like Bitcoin and Ethereum in India, considering their potential impact on the economy, financial stability, and security concerns.

Artificial Intelligence and Job Displacement: Exploring the implications of artificial intelligence (AI) and automation on employment patterns, workforce reskilling, and the role of governments in managing job displacement.

Farm Bills and Agricultural Reforms: Analyzing the controversial farm bills passed by the Indian government, their potential impact on farmers, agricultural markets, and food security, and the ongoing protests against them.

One Nation One Election: Debating the proposal for simultaneous elections to the Lok Sabha and state legislative assemblies, discussing its feasibility, advantages, and challenges in implementation.

Right to Privacy vs. National Security: Discussing the balance between individual privacy rights and national security interests, especially in the context of surveillance, data collection, and government policies.

New Education Policy (NEP): Evaluating the key provisions and objectives of the New Education Policy 2020, its potential to transform the Indian education system, and challenges in its implementation.

India-China Border Dispute: Analyzing the border tensions between India and China, diplomatic efforts to resolve the standoff, and implications for regional security and geopolitics.

COVID-19 Vaccination Strategy: Assessing the rollout of COVID-19 vaccines in India, challenges in distribution, vaccine hesitancy, and strategies to achieve herd immunity and control the pandemic.

Privatization of Public Sector Undertakings (PSUs): Debating the government's decision to privatize certain PSUs, its impact on employment, economic efficiency, and the role of PSUs in the Indian economy.

Startup Ecosystem in India: Exploring the growth of startups in India, government initiatives to promote entrepreneurship, and challenges faced by startups in funding, regulatory compliance, and market access.

Ease of Doing Business in India: Analyzing India's ranking in the Ease of Doing Business Index, regulatory reforms to attract foreign investment, and challenges in improving the business environment.

5G Technology Deployment: Discussing the potential of 5G technology in India, its applications in sectors like telecommunications, healthcare, and smart cities, and challenges in infrastructure development and security.

Universal Basic Income (UBI): Exploring the concept of UBI as a social welfare scheme, its feasibility, benefits, and implications for poverty alleviation, income inequality, and fiscal sustainability.

Electric Vehicles (EVs) Adoption: Analyzing the government's push for electric vehicles, challenges in EV adoption, such as infrastructure development, affordability, and environmental benefits.

Remote Work Culture: Discuss the shift towards remote work due to the COVID-19 pandemic, its impact on productivity, work-life balance, and the future of traditional office spaces.

Tourism Industry Revival: Assessing strategies to revive the tourism industry post-pandemic, promoting domestic tourism, infrastructure development, and sustainable tourism practices.

Artificial Intelligence (AI) in Healthcare: Discuss the role of AI in revolutionizing healthcare delivery, diagnosis, treatment, and patient care.

Blockchain Technology and Its Applications: Explore the potential of blockchain technology beyond cryptocurrencies, such as supply chain management, digital identity verification, and secure transactions.

Internet of Things (IoT) and Smart Cities: Analyze how IoT technology can enhance urban infrastructure, resource management, public services, and citizen engagement in smart city initiatives.

Big Data Analytics in Business Decision Making: Discuss the use of big data analytics in industries like finance, marketing, healthcare, and manufacturing to derive actionable insights and drive strategic decision-making.

Renewable Energy Integration into the Power Grid: Examine the challenges and solutions for integrating renewable energy sources like solar and wind into the existing power grid to achieve a sustainable energy future.

Cybersecurity Threats and Countermeasures: Explore the evolving landscape of cybersecurity threats, including malware, phishing attacks, and data breaches, and strategies to enhance cybersecurity posture in organizations and governments.

Future of Transportation: Electric Vehicles and Autonomous Vehicles: Discuss the adoption of electric vehicles (EVs) and autonomous vehicles (AVs), their impact on urban mobility, infrastructure requirements, and regulatory challenges.

Augmented Reality (AR) and Virtual Reality (VR) in Education: Evaluate the potential of AR and VR technologies in transforming the education sector, enhancing learning experiences, and improving student engagement.

Space Exploration and Colonization: Debate the significance of space exploration missions, potential benefits of space colonization, and ethical considerations related to human expansion beyond Earth.

Biotechnology Innovations in Agriculture: Discuss the role of biotechnology in agriculture, including genetically modified organisms (GMOs), gene editing techniques, and bio-fortification to address global food security challenges.

5G Technology and Its Implications: Analyze the impact of 5G technology on communication networks, Internet of Things (IoT) devices, mobile computing, and emerging applications like remote surgery and autonomous vehicles.

Future of Work: Automation and Job Displacement: Explore the implications of automation, robotics, and artificial intelligence on the future of work, workforce reskilling, and socio-economic dynamics.

Green Building Technologies and Sustainable Architecture: Discuss innovative green building technologies, sustainable design practices, and the importance of energy-efficient buildings in mitigating climate change.

Nanotechnology: Applications and Challenges: Evaluate the potential applications of nanotechnology in medicine, electronics, materials science, and environmental remediation, along with safety and ethical concerns.

Quantum Computing: Opportunities and Limitations: Examine the potential of quantum computing to revolutionize computation, cryptography, and scientific research, as well as the challenges in achieving practical quantum systems.

Impact of Climate Change on Global Weather Patterns: Discuss the observed changes in weather patterns worldwide due to climate change, including extreme weather events, heat waves, droughts, and floods.

Role of Deforestation in Climate Change: Explore the contribution of deforestation to greenhouse gas emissions, loss of biodiversity, and disruptions to ecosystems, and discuss strategies for sustainable forest management and reforestation.

Renewable Energy Transition: Challenges and Opportunities: Analyze the transition from fossil fuels to renewable energy sources like solar, wind, and hydroelectric power, and discuss policy measures to accelerate the shift towards clean energy.

Ocean Acidification and Its Impact on Marine Ecosystems: Examine the causes and consequences of ocean acidification resulting from increased carbon dioxide absorption, and discuss measures to mitigate its effects on coral reefs, marine life, and coastal communities.

Plastic Pollution: Addressing the Plastic Crisis: Discuss the environmental and human health impacts of plastic pollution in oceans, rivers, and landfills, and explore solutions such as plastic recycling, bans on single-use plastics, and innovative biodegradable alternatives.

Urban Air Pollution and Public Health: Evaluate the sources and effects of air pollution in urban areas, including emissions from vehicles, industries, and biomass burning, and discuss policies and technologies to improve air quality and protect public health.

Adaptation Strategies for Climate Resilience: Examine strategies for building resilience to climate change impacts, including infrastructure upgrades, early warning systems, disaster preparedness, and community-based adaptation initiatives.

Biodiversity Conservation and Habitat Restoration: Discuss the importance of preserving biodiversity hotspots, protecting endangered species, and restoring degraded habitats to enhance ecosystem resilience and ecological balance.

Sustainable Agriculture Practices: Explore sustainable farming techniques such as organic farming, agroforestry, and precision agriculture to minimize environmental impacts, conserve soil and water resources, and enhance food security.

Carbon Pricing and Market Mechanisms: Analyze the effectiveness of carbon pricing mechanisms such as carbon taxes and cap-and-trade systems in reducing greenhouse gas emissions and incentivizing low-carbon investments.

Role of International Cooperation in Climate Action: Discuss the significance of international agreements like the Paris Agreement in combating climate change, addressing climate finance, technology transfer, and capacity building, and promoting global solidarity in climate action.

Impacts of Climate Change on Indigenous Communities: Explore the unique vulnerabilities of indigenous communities to climate change impacts such as land loss, displacement, and loss of traditional livelihoods, and discuss strategies for supporting indigenous resilience and adaptation.

Eco-friendly Transportation Solutions: Evaluate sustainable transportation options such as public transit, cycling infrastructure, electric vehicles, and green urban planning to reduce carbon emissions and improve air quality in cities.

Role of Youth in Climate Activism: Discuss the role of youth-led movements and activism in raising awareness about climate change, demanding climate action from policymakers, and driving grassroots initiatives for environmental sustainability.

Green Technologies for Climate Mitigation: Explore innovative technologies such as carbon capture and storage (CCS), renewable energy storage, and sustainable materials for building a low-carbon economy and achieving net-zero emissions targets.

Renewable Energy Targets and Policy Framework: Discuss India's targets for renewable energy capacity expansion, the effectiveness of policies such as the National Solar Mission and National Wind Mission, and challenges in achieving renewable energy integration into the grid.

Energy Transition: Balancing Coal and Renewable Energy: Analyze the transition from coal-based power generation to renewable energy sources, including the implications for energy security, environmental sustainability, and economic growth.

Electricity Distribution Reforms: Explore reforms in electricity distribution aimed at improving efficiency, reducing losses, promoting competition, and enhancing consumer satisfaction, such as privatization, franchise models, and smart grid initiatives.

Energy Access and Rural Electrification: Discuss initiatives like the Saubhagya Scheme aimed at providing electricity access to all households, challenges in last-mile connectivity, and innovative solutions for off-grid and decentralized energy solutions in rural areas.

Grid Modernization and Smart Grid Technologies: Examine the role of smart grid technologies in enhancing grid reliability, efficiency, and resilience, including advanced metering infrastructure, demand response systems, and grid automation.

Power Sector Financial Health: Analyze the financial health of power sector utilities, including issues such as mounting debt, revenue losses, and subsidy burden, and discuss strategies for financial restructuring, tariff reforms, and improving operational efficiency.

Coal Sector Reforms and Clean Coal Technologies: Explore reforms in the coal sector aimed at enhancing efficiency, environmental sustainability, and coal beneficiation technologies, including the adoption of clean coal technologies such as supercritical and ultra-supercritical power plants.

Energy Storage Solutions: Discuss the role of energy storage technologies such as battery storage, pumped hydro storage, and thermal energy storage in integrating renewable energy, enhancing grid stability, and meeting peak demand requirements.

Electric Vehicle (EV) Integration and Charging Infrastructure: Examine policies and initiatives for promoting electric mobility, challenges in EV adoption, and the development of EV charging infrastructure to support the transition to electric vehicles.

Cross-Border Energy Trade and Regional Cooperation: Explore opportunities for cross-border energy trade, including electricity import/export agreements, interconnections with neighboring countries, and regional cooperation initiatives for energy security and economic development.

Decentralized Energy Generation and Microgrids: Discuss the role of decentralized energy generation, including rooftop solar, microgrids, and community-based renewable energy projects, in enhancing energy access, resilience, and sustainability.

Nuclear Energy Expansion and Safety: Evaluate the prospects for nuclear energy expansion in India, including the development of new nuclear power plants, nuclear safety regulations, and public perception of nuclear energy post-Fukushima.

Energy Efficiency and Demand-Side Management: Analyze strategies for promoting energy efficiency and demand-side management, including energy conservation measures, energy-efficient appliances, and building energy codes.

Role of Public-Private Partnerships (PPPs) in Power Sector Development: Discuss the role of PPPs in financing, developing, and operating power sector infrastructure, including generation, transmission, and distribution projects.

Climate Change Mitigation and Adaptation in the Power Sector: Explore measures for climate change mitigation and adaptation in the power sector, including emission reduction strategies, climate-resilient infrastructure, and carbon pricing mechanisms.

Group discussions are often used as a way to grade students on competitive tests, interviews, and other selection processes. To do well in a group talk, you need to be able to communicate well, know a lot about the topic, and be able to work with others. Here are some tips on how to get high scores in a normal group discussion and how the points are given out:

How to score more points in a Group Discussion:

1. Knowledge of the content (30–40%):

— Making it clear that they understand the subject.
– Giving accurate and useful information.
— Being aware of present events and settings.

2. Talking and writing skills (25–30%):
– Speaking clearly and with good grammar.
— Going at a modest speed and tone.
-- Using the right words and wording.

3. Logical and Analytical Ability (20–25%):
– Critically analyzing the subject.
   - Making a case that makes sense and is based on logic.
— Making claims that are true and well-supported.

4. People Interpersonal skills (15–20%):
— Listening to other people.
– Respecting different points of view.
– Making a useful contribution to the conversation.

5. Strength and initiative (10–15%):
- Starting points for conversation.
   - Making sure the talk stays on track.
– Driving the talk forward by taking the lead.

Here are some tips to do well in a group discussion:

1. Read and comprehend the subject: Fully read and comprehend the subject.
- Clear up any questions before the talk starts.

2. Active participation means adding important points to the conversation.
- Try not to be too powerful or too weak.

3. "Quality over Quantity": Don't worry about how much you talk; instead, focus on making good comments.
- Make sure that every point you make adds something to the conversation.

4. Listen Actively: This means to hear what other people are saying.

—Acknowledge other people's points and build on them.

5. "Maintain Clarity": Talk clearly and concisely.
- Get your thoughts in order before you say them.

6. Stay Calm and Composed: - Stay calm even when you disagree with someone.
- Try not to get angry or defensive.

7. Be Open to Feedback: - Be willing to take constructive feedback with a smile.
— Change how you do things based on what people say.

8. Make sure that everyone has a chance to speak by "balancing the conversation."
—Encourage people who are quiet to say what they think.

9. Show leadership by starting a conversation or making a new point.
- Lead the conversation to make sure it stays on track.

10. Be Respectful: - Value different points of view, even if you don't agree with them.
- Don't talk over other people when they're talking.

11. Time Management: - Pay attention to the time allotted for the talk.
- Make sure the group talks about more than one part of the subject.

12. Body Language: - Keep your body language upbeat and sure of yourself.
- Don't look like you're not interested or are busy.

13. Give examples and make comparisons:
—Back up your points with important examples.
- Use comparisons to make hard ideas easier to understand.

14. Stay Up to Date:
- Keep up with current events and important issues.
- Use recent examples to back up what you're saying.

15. Practice Group talks: — Have practice group talks to get better at them.
- Ask your friends or a mentor for feedback.

Remember that the point is not only to say what you think but also to add to a talk that is constructive and works well with others. You can improve your chances of doing well in a group discussion by focusing on your understanding of the topic, your ability to communicate, and your ability to work with others.

CHAPTER FIVE

Cracking the Interview: How to Do Well in an Interview

Here are common mistakes categorized into three stages of the interview process: pre-interview, during the interview, and post-interview.

Pre-Interview Mistakes:

1. Lack of Research:

- Mistake: Failing to thoroughly research the company, its culture, and its role.

- Impact: You may appear disinterested or unprepared during the interview.

2. Ignoring the Job Description:

- Mistake: Not aligning your resume and preparation with the requirements outlined in the job description.

- Impact: Your responses may not address the specific needs of the role.

3. Inadequate Preparation:

- Mistake: Not practicing responses to common interview questions or preparing for behavioral scenarios.

- Impact: You may struggle to articulate your experiences and skills during the interview.

4. Poor Time Management:

- Mistake: Failing to plan your route or underestimating travel time.

- Impact: Arriving late may create a negative first impression.

During the Interview Mistakes:

5. Ineffective Communication:

- Mistake: Speaking too fast, using unclear language, or providing overly lengthy responses.

- Impact: It can hinder the interviewer's understanding of your points.

6. Lack of Confidence:

- Mistake: Displaying nervousness, lack of eye contact, or weak body language.

- Impact: It may convey a lack of confidence in your abilities.

7. Overlooking Non-Verbal Cues:

- Mistake: Ignoring the interviewer's body language or failing to adjust your own.

- Impact: Misreading cues can lead to misunderstandings or miscommunication.

8. Talking Over Others:

- Mistake: Interrupting the interviewer or not allowing them to finish questions.

- Impact: It can be perceived as disrespectful and hinder effective communication.

Post-Interview Mistakes:

9. Neglecting Follow-Up:

- Mistake: Failing to send a thank-you email after the interview.

- Impact: It may be interpreted as a lack of appreciation for the opportunity.

10. Delaying Reflection:

- Mistake: Not reflecting on the interview and areas for improvement.

- Impact: You may repeat mistakes in future interviews.

11. Overlooking Feedback:

- Mistake: Dismissing feedback from mock interviews or interview coaches.

- Impact: Missed opportunities for self-improvement.

12. Not Updating References:

- Mistake: Assuming your references are current without verifying.

- Impact: Outdated references may not provide relevant insights.

These categorized mistakes cover key aspects of the interview process. Addressing these areas can significantly enhance your chances of presenting yourself in the best possible light throughout the entire interview journey.

Pre-interview activities are crucial for preparing yourself mentally and physically for the interview.

Here's a comprehensive list of pre-interview activities every candidate should plan.

1. Research the Company:

- Understand the company's mission, values, culture, products/services, recent news, and industry reputation.

- Explore their website, social media profiles, annual reports, and news articles.

2. Understand the Role:

- Review the job description thoroughly to understand the responsibilities, qualifications, and skills required.

- Identify how your experiences, skills, and strengths align with the requirements of the role.

3. Prepare Your Resume and Cover Letter:

- Tailor your resume and cover letter to highlight relevant experiences, achievements, and skills.

- Ensure your documents are error-free, well-formatted, and customized for the specific job.

4. Practice Common Interview Questions:

- Prepare responses to common interview questions, including behavioral, situational, and technical questions.

- Use the Mock Situation analysis for every question, with a counter question.

5. Research the Interview Format:

- Understand the format of the interview (e.g., one-on-one, panel, behavioral interview).

- Prepare accordingly, considering the type of questions and interactions you may encounter.

6. Plan Your Outfit:

- Choose professional attire that is appropriate for the company culture and industry.

- Ensure your outfit is clean, well-fitted, and makes you feel confident.

7. Practice Good Body Language:

- Practice maintaining good posture, making eye contact, and offering a firm handshake.

- Pay attention to your facial expressions and non-verbal cues.

8. Plan Your Route:

- Research the interview location and plan your route.
- Consider factors such as traffic, parking availability, public transportation options, and estimated travel time.

9. Gather Necessary Documents:
- Prepare copies of your resume, cover letter, references, portfolio, and any other relevant documents.
- Organize them neatly in a folder or portfolio to bring to the interview.

10. Review Your Online Presence:
- Ensure your LinkedIn profile and other professional social media accounts are up-to-date and present a positive image.
- Google your name to see what information comes up and address any red flags if necessary.

11. Get Plenty of Rest:
- Aim to get a good night's sleep before the interview to feel refreshed and alert.
- Avoid staying up late or engaging in activities that may affect your sleep quality.

12. Practice Self-Care:
- Take time to relax and engage in activities that help you de-stress and feel confident.
- Practice deep breathing exercises, meditation, or mindfulness techniques if needed.

By planning and executing these pre-interview activities effectively, you'll be better prepared to present yourself confidently and professionally during the interview process.

Avoiding bad body language during an interview is crucial as it contributes significantly to the overall impression you make on the interviewer. Here are tips to help you steer clear of negative body language:

1. Maintain Good Posture:
- Bad Body Language: Slouching or leaning back excessively.
- Avoidance Tip: Sit up straight with your back against the chair. Maintain an open and engaged posture.

2. Eye Contact:
- Bad Body Language: Avoiding eye contact can make you appear disinterested or unconfident.
- Avoidance Tip: Make regular, natural eye contact to convey confidence and engagement. Avoid staring but maintain a friendly gaze.

3. Weak Handshake:

- Bad Body Language: Offering a limp or overly strong handshake.
- Avoidance Tip: Aim for a firm handshake, matching the pressure of the other person. It should be confident, but not overpowering.

4. Excessive Fidgeting:
- Bad Body Language: Tapping your foot, playing with your hair, or excessive hand movements can be distracting.
- Avoidance Tip: Be aware of your movements and keep them to a minimum. Cross your legs at the ankles, not the knees, to appear more composed.

5. Crossing Arms:
- Bad Body Language: Crossing your arms can signal defensiveness or closed-mindedness.
- Avoidance Tip: Keep your arms open and relaxed. Resting your hands on your lap or the armrests of the chair can convey openness.

6. Nervous Habits:
- Bad Body Language: Nail biting, hair twirling, or other nervous habits.
- Avoidance Tip: Be mindful of your tendencies and consciously avoid engaging in nervous habits during the interview.

7. Facial Expressions:
- Bad Body Language: Frowning, scowling, or expressions of disapproval.
- Avoidance Tip: Maintain a pleasant and neutral facial expression. Smile naturally when appropriate to convey warmth.

8. Too Much or Too Little Gesturing:
- Bad Body Language: Excessive gesturing can be distracting, while too little can make you appear unengaged.
- Avoidance Tip: Use natural gestures to emphasize points but avoid overdoing it. Be mindful of your hand movements.

9. Poor Spatial Awareness:
- Bad Body Language: Invading personal space or sitting too far away can make others uncomfortable.
- Avoidance Tip: Respect personal space. Sit at a comfortable distance, neither too close nor too far from the interviewer.

10. Lack of Smile:
- Bad Body Language: Not smiling can make you appear unfriendly or disinterested.
- Avoidance Tip: Smile naturally during appropriate moments, such as when greeting the interviewer or expressing enthusiasm.

11. Distraction with Personal Items:

- Bad Body Language: Checking your phone or fidgeting with personal items.

- Avoidance Tip: Keep your phone on silent and out of sight. Focus on the interview without distractions.

12. Lack of Engagement:

- Bad Body Language: Zoning out or looking disinterested.

- Avoidance Tip: Stay engaged, nod occasionally to show you're listening, and respond actively to the conversation.

Being aware of your body language and making conscious efforts to convey professionalism and positivity will contribute to a more successful interview. Practice in front of a mirror or with a friend to receive feedback on your non-verbal cues.

Post-interview communication with HR is an important aspect of the interview process. It allows you to express gratitude, reiterate your interest in the position, and leave a positive impression. Here are some good ways to communicate with HR after an interview:

1. Send a Thank-You Email:

- Timing: Send the thank-you email within 24 hours of the interview.

- Content:

- Express gratitude for the opportunity to interview.

- Reiterate your enthusiasm for the position and the company.

- Mention a specific aspect of the interview that you found interesting or valuable.

- Highlight one or two key reasons why you believe you are a strong fit for the role.

- Express your eagerness to move forward in the hiring process.

2. Personalize Your Message:

- Reference Specifics: Mention specific topics discussed during the interview to show that you were actively engaged.

- Personal Touch: If you connected with the interviewer on a personal or shared interest, briefly reference it in your thank-you email.

3. Clarify Any Points:

- Follow-Up Questions: If there were any points discussed during the interview that need clarification or additional information, politely ask for clarification in your email.

- Reiterate Interest: Frame your questions in a way that shows your continued interest in the role and eagerness to contribute.

4. Reaffirm Your Fit:

- Reinforce Your Fit: Reiterate how your skills, experiences, and qualifications align with the requirements of the role.

- Match with Company Culture: Emphasize how your values and work style align with the company culture.

5. Professional Tone:

- Formal Language: Maintain a professional and formal tone in your communication.

- Proofread: Ensure your email is free of grammatical errors and typos.

6. Express Flexibility:

- Availability: If applicable, express your flexibility regarding follow-up interviews or additional assessments.

- Availability for Questions: Mention that you are available for any additional information or questions they may have.

7. Use the Right Subject Line:

- Clear Subject Line: Make your subject line clear and specific, such as "Thank You for the Interview – [Your Full Name]."

- Reference the Position: Include the position title in the subject line.

8. LinkedIn Connection:

- Connect on LinkedIn: If you haven't already, consider sending a LinkedIn connection request.

- Express Appreciation: In your connection message, express appreciation for the opportunity and mention your continued interest.

9. Follow Company Protocol:

- Follow Instructions: If the company provided specific instructions on post-interview communication, ensure you follow them.

- Use Preferred Method: If HR indicated a preferred method of communication, adhere to their preferences.

10. Maintain Professionalism:

- Avoid Desperation: While expressing enthusiasm, avoid sounding desperate for the job.

- Professional Language: Keep your language professional and avoid using overly casual or informal expressions.

Remember, effective post-interview communication reflects your professionalism, attention to detail, and genuine interest in the position. It's an opportunity to stand out and leave a positive lasting impression.

Best practices and common sense can help you prepare for an interview. Here is a list of tips that are in line with good interview strategies:

1. Get to Know Yourself: "Know your strengths, weaknesses, and what makes you special before you go into the interview room." Being aware of yourself is the key to doing well in an interview.

2. Learn about the company: "Employers like applicants who have taken the time to learn about their values, mission, and most recent accomplishments." Your interest in the company is clear from how much you know about it.

3. Practice, Practice, Practice: "Go over your answers to common questions again and again." This makes you feel better about yourself and helps you say what you want to say in the interview.

4. Show Off Your Accomplishments: "Give specific examples of what you've done well." Whenever you can, put a number on your accomplishments. This gives your claims more weight.

5. Show that you can adapt: "Show that you can adapt by talking about problems you've had and how you solved them." Employers want to hire people who can do well in changing situations.

6. Ask Deep Questions: "Think of deep questions you could ask about the job, the team, or the company." This shows that you are genuinely interested in and committed to the possible chance.

7. Show Off Your Soft Skills: "In addition to your technical skills, you should highlight your soft skills, like how well you can communicate, work with others, and solve problems." Often, these traits are just as important as professional know-how.

8. Stay Positive: "Take a positive view of the interview the whole time." When you talk about problems, keep your attention on the answers and what you learned from them.

9. Use the Behavioral Questions: "Use the behavioral questions." with the help of Question related This answer is clear and to the point.

10. Do something after the interview: "Send a thank-you email and reiterate your interest in the job." This shows that you're grateful and keeps the interviewer think about you.

Remember that these tips are general ways to do well in an interview, but you should still make sure that your approach fits the business, role, and company culture. Every interview is different, and showing that you are real and excited about the job can help you stand out.

A Story of Failing and Rising Above in the Interview Room

Starting off:

A young man named Arjun lived in the busy city of Mumbai, where there were tall buildings and lots of honking horns. Arjun's hopes and goals were as big as the Arabian Sea, which was next to the city, and as high as the planes taking off from Chhatrapati Shivaji Maharaj International Airport. What was his main goal? To get a job at one of India's best global companies, which is something everyone wants. However, Arjun had no idea that his path to success would be full of surprises. It all started with a fateful interview that would change his life forever.

The Meeting:

Arjun had carefully studied for the interview at one of the best tech companies in India. He spent a lot of time learning about the company, practicing answers to questions that might be asked, and making his resume look its best. He walked into the interview room full of confidence and wearing his best suit. He felt good about himself.

When the interview started, Arjun started a lively chat with the people who were interviewing him. He talked about his experiences beautifully, showed off his technical skills, and answered every question with confidence. But Arjun didn't know it, but his body language showed that he was having a hard time.

What went wrong:

As the conversation went on, the people interviewing Arjun could tell that he was a little nervous. His hands shook a little when he made gestures, his shoulders tensed up every once in a while, and he lost his focus on the person he was talking to at key points. Even though Arjun's answers were amazing, his body language showed that he was nervous and insecure.

Arjun's downfall was his lack of knowledge about his body language, even though he was qualified and skilled. At the end of the interview, both interviewers gave each other knowing looks, as if they were aware of how the candidate's body language had affected their opinion of them.

The Turn Down:

After a few weeks, Arjun was eagerly waiting to hear how the interview went. That being said, when the email finally came, it was a terrible blow: his application had been turned down. Arjun was devastated and confused, and he had no idea what he had done wrong. Didn't he do great on the hard questions? Hadn't he shown how much he wanted the part? When he thought back on his interview, he realized the truth: his otherwise perfect qualities had been ruined by the small changes in his body language.

What I Thought:

Arjun set out on a journey to learn about himself and make himself better because he was determined to do so. It was important to him to get feedback from teachers, take communication workshops, and learn more about the psychology of body language. Arjun found that his bad body language was caused by fear of failing, self-doubt, and a lack of confidence. He did this by reflecting on himself and not giving up.

With this new knowledge, Arjun set out to change his body language from something that was hurting him to something that was helping him. He did power poses to show confidence, mirroring techniques to get to know people, and learning how to make and keep strong eye contact. Arjun felt more sure of himself, more confident, and more ready to take on the interview room with each passing day.

The Coming Back:

With the new information and confidence he had gained, Arjun set out to make things right with possible employers. He went back to the competitive world of job interviews with a fresh sense of purpose and an up-to-date resume. Now, though, he went into each interview with more confidence and awareness than ever before.

Arjun was so sure of himself as he sat across from the interviews. It said a lot about him that he stood straight, made deliberate movements, and kept eye contact the whole time. With every word he spoke, Arjun held the room's attention with a magnetic personality that held the interviewers' attention.

In the weeks that followed, Arjun got a lot of job interview offers from the best companies in the country. The skills he had and the way he had learned to read body language both impressed employers. Each interview that went well made Arjun feel even better about himself, and his dreams of having a successful job in business became a reality.

In conclusion:

An important thing that Arjun's trip makes me remember is how important body language is during an interview. Learning how to use body language well is important for success in a world where first impressions can make or break a candidate's chances. Arjun turned his first failure into a springboard for growth and success by not giving up, reflecting on himself, and making a promise to keep getting better.

As Arjun looks ahead, he knows that the lessons he has learned on his trip will help him in every part of his life. He's ready to take on any task that comes his way now that he has more confidence, poise, and an understanding of how powerful body language can be. Failure isn't something you want to avoid in life; you want to welcome it, learn from it, and come out better on the other side. The best is yet to come for Arjun.

CHAPTER SIX

BUILDING A 30-DAYS : ACTION PLAN

Staying committed to your personal and professional development journey can sometimes be challenging, but it's essential for growth and success. Here are some ways to stay encouraged and motivated along the way:

1. Set Clear Goals: Define specific, measurable goals for your journey. Having clear objectives will give you direction and purpose, making it easier to stay motivated.

2. Celebrate Small Wins: Acknowledge and celebrate every achievement, no matter how small. Each step forward is progress, and recognizing your accomplishments will boost your morale and motivation.

3. Visualize Success: Create a mental image of what success looks like for you. Visualizing your goals and imagining yourself achieving them can be a powerful motivator, helping you stay focused and determined.

4. Stay Positive: Maintain a positive mindset even in the face of challenges. Practice gratitude, affirmations, and positive self-talk to cultivate optimism and resilience.

5. Surround Yourself with Support: Surround yourself with supportive friends, family members, mentors, and peers who believe in you and your journey. Their encouragement and words of affirmation can fuel your motivation during tough times.

6. Find Inspiration: Seek out inspiring stories, quotes, books, podcasts, or videos that resonate with your goals and aspirations. Drawing inspiration from others' success stories can reignite your passion and drive.

7. Create Accountability: Share your goals with someone you trust and ask them to hold you accountable. Regular check-ins and progress updates will keep you accountable and motivated to stay on track.

8. Break it Down: Break your journey into smaller, manageable tasks or milestones. Focus on completing one task at a time, and celebrate your progress along the way.

9. Embrace Challenges: View challenges as opportunities for growth rather than obstacles. Embrace discomfort and push yourself out of your comfort zone, knowing that overcoming challenges will make you stronger and more resilient.

10. Reflect on Your Why: Remind yourself of why you embarked on this journey in the first place. Reconnect with your passion, purpose, and values to reignite your motivation and commitment.

11. Practice Self-Care: Take care of your physical, mental, and emotional well-being. Prioritize self-care activities such as exercise, meditation, hobbies, and downtime to recharge and rejuvenate.

12. Stay Flexible: Be open to adjusting your plans and strategies as needed. Flexibility is key to navigating obstacles and setbacks effectively while staying focused on your long-term goals.

By incorporating these strategies into your journey, you'll be better equipped to stay encouraged, motivated, and committed as you work towards achieving your goals and aspirations. Remember, the journey may have its ups and downs, but every step forward brings you closer to success.

Dhirubhai Ambani's journey from a small village in Gujarat to founding one of India's largest conglomerates, Reliance Industries, is a true inspiration for entrepreneurs worldwide. Here's a motivational story of Dhirubhai Ambani and his relentless pursuit of success:

Humble Beginnings:

Dhirubhai Ambani was born into a modest family in Chorwad, Gujarat, in 1932. From a young age, he displayed an entrepreneurial spirit, selling snacks to pilgrims visiting Mount Girnar. Despite facing financial constraints, Dhirubhai possessed a vision far beyond his circumstances.

Ambition and Vision:

Dhirubhai harbored a grand vision to revolutionize India's business landscape and empower millions of ordinary Indians. He believed in the power of entrepreneurship to create wealth and opportunities for all, regardless of background or social status.

Relentless Determination:

In 1958, Dhirubhai moved to Yemen to work as a clerk in a trading firm. During his time there, he honed his business acumen and learned the intricacies of commodity trading. Despite facing numerous challenges and setbacks, Dhirubhai remained undeterred in his pursuit of success.

The Birth of Reliance:

In 1966, Dhirubhai returned to India with a clear vision to build his own business empire. With just a small capital and unwavering determination, he founded Reliance Commercial Corporation, laying the foundation for what would later become Reliance Industries Limited.

Unconventional Approach:

Dhirubhai's approach to business was unconventional yet visionary. He leveraged innovative financial strategies, including equity financing and aggressive expansion, to fuel Reliance's growth. He believed in taking calculated risks and was not afraid to challenge the status quo.

Transforming Industries:

Under Dhirubhai's leadership, Reliance ventured into diverse sectors, including textiles, petrochemicals, telecommunications, and energy. His bold initiatives and relentless pursuit of excellence transformed traditional industries, driving innovation and creating unprecedented value.

Empowering the Masses:

Dhirubhai's greatest legacy lies in his commitment to empowering ordinary Indians. He democratized wealth creation by offering shares of Reliance to the public through initial public offerings (IPOs), allowing millions of Indians to become shareholders and participate in India's economic growth.

Resilience in Adversity:

Dhirubhai faced numerous challenges throughout his entrepreneurial journey, including regulatory hurdles, business rivalries, and personal attacks. However, he remained resilient in the face of adversity, turning obstacles into opportunities and emerging stronger than ever.

Legacy of Inspiration:

Dhirubhai Ambani's entrepreneurial journey continues to inspire generations of entrepreneurs worldwide. His relentless ambition, unwavering determination, and pioneering spirit exemplify the limitless possibilities of the human spirit and serve as a beacon of hope for aspiring business leaders.

Conclusion:

Dhirubhai Ambani's life is a testament to the power of vision, determination, and perseverance. From humble beginnings to building a global business empire, his journey embodies the true essence of entrepreneurship and serves as a source of inspiration for individuals striving to achieve their dreams against all odds.

Building a 30-Day Career Development Action Plan: Week-by-Week Guide

We made a sample action plan, but each one was different based on the candidate's level of experience, the job they were applying for, and the company they wanted to work for.

Week 1: Self-Discovery and Assessment

Reflect on Goals and Values

Day 1: ____/_____/_______ (DD / MM / YYYY) – 4 Hours

Assignment

1. Write About your career goals (In 500 words)

Ideal Job Titles | Company Aims. | Ideals Job Location | Company Address | Website

1.
2.
3.
4.
5.
6.
7.
8.
9.
10.

Selection Based : Identify three key priorities that are non-negotiable in your career from the below options.

{Location | Salary | Company Name (Brand) | Growth & Learning | Job Security | Job Title}

Priority 1

Priority 2

Priority 3

Skills Assessment

Day 2: ____/_____/_______ (DD / MM / YYYY) – 4 Hours

- **Assignment:** List your top skills and identify areas for improvement

Ranking | Skill Set | Rate your skills (Excellent =5 and Poor = 1)

1

2

3

4

5

6

7

8

9

10

Test your Skills from various available websites

1. Shine Learning: Shine Learning offers a variety of free skill assessment tests covering topics such as aptitude, technical skills, and personality assessments.

- Website: https://www.shine.com/learning/free-assessment-tests

2. MyPerfectResume: MyPerfectResume provides free skills assessment tests tailored for job seekers, covering areas such as soft skills, technical skills, and industry-specific skills.

- Website: https://www.myperfectresume.com/career-center/skills-assessment

3. TalentSprint: TalentSprint offers free online skill assessment tests in collaboration with industry experts, covering topics such as coding, data science, and digital marketing.

- Website: https://www.talentsprint.com/assessments/

4. IndiaBIX: IndiaBIX offers a wide range of free aptitude tests, technical tests, and interview preparation resources to help job seekers assess and improve their skills.

- Website: https://www.indiabix.com/online-test/

5. Youth4Work: Youth4Work provides free skill assessment tests and certification exams across various domains, including programming, finance, and communication skills.

- Website: https://www.prep.youth4work.com/

6. Mettl: Mettl offers free skill assessment tests and quizzes covering a broad range of topics, including cognitive skills, behavioral skills, and domain-specific skills.

- Website: https://mettl.com/free-practice-tests/

7. Testbook: Testbook provides free practice tests and quizzes for competitive exams, government job exams, and skill-based assessments, helping candidates prepare for various recruitment tests.

- Website: https://testbook.com/practice

8. Careers360: Careers360 offers free online mock tests and skill assessment tools for competitive exams, entrance tests, and job recruitment exams.

- Website: https://www.careers360.com/online-exam

9. India Skills: India Skills offers free skill assessment tests and quizzes across multiple domains, helping individuals evaluate their competencies and identify areas for improvement.

- Website: https://www.indiaskills.edu.in/

10. Freshersworld: Freshersworld provides free skill assessment tests and quizzes for fresh graduates and job seekers, covering areas such as aptitude, reasoning, and technical skills.

- Website: https://www.freshersworld.com/assessment-tests/

These websites offer a wide range of skill assessment tests and resources to help individuals evaluate their strengths, identify areas for improvement, and enhance their employability in the Indian job market.

Personality and Work Style

Day 3: _____/_____/_______ (DD / MM / YYYY) – 4 Hours

- Assignment: Take a personality assessment

Here are ten websites offering reliable personality assessments suitable for individuals in India:

1. 16Personalities:

- Website: https://www.16personalities.com/

- Description: Offers a free personality test based on the Myers-Briggs Type Indicator (MBTI), providing insights into personality types and traits.

2. Truity:

- Website: https://www.truity.com/

- Description: Provides a range of personality tests, including the Myers-Briggs Type Indicator (MBTI), the Big Five Personality Test, and career-specific assessments.

3. Psych Central:

- Website: https://psychcentral.com/quizzes/personality/

- Description: Offers a collection of free personality quizzes covering various aspects of personality, behavior, and mental health.

4. HumanMetrics:

- Website: https://www.humanmetrics.com/personality

- Description: Provides free personality tests based on the Myers-Briggs Type Indicator (MBTI), offering insights into personality preferences and characteristics.

5. Similar Minds:
- Website: http://similarminds.com/
- Description: Offers a variety of free personality tests, including the Jung Typology Test and Enneagram Test, to help individuals understand their personality traits and preferences.

6. MyPersonality.Info:
- Website: https://mypersonality.info/
- Description: Provides free personality tests based on the Big Five personality traits, offering insights into individual personality dimensions and characteristics.

7. PsychTests:
- Website: https://www.psychtests.com/
- Description: Offers a range of scientifically validated personality tests and assessments covering various aspects of personality, behavior, and emotional intelligence.

8. Online Personality Tests:
- Website: https://www.onlinepersonalitytests.org/
- Description: Provides free personality tests and quizzes, including the Myers-Briggs Type Indicator (MBTI), the Big Five Personality Test, and career-oriented assessments.

9. 123Test:
- Website: https://www.123test.com/personality-test/
- Description: Offers a free personality test based on the Big Five personality traits, providing insights into individual personality dimensions and characteristics.

10. Psychology Today:
- Website: https://www.psychologytoday.com/us/tests/personality
- Description: Provides a variety of free personality tests and quizzes developed by psychologists and experts in the field, covering different aspects of personality and behavior.

These websites offer a range of personality assessments that can provide valuable insights into individual traits, preferences, and behaviors, helping individuals gain self-awareness and understand their strengths and areas for development.

Reflect on how your personality aligns with different career paths, now rework on Day 1 ,2 and 3 t verify that do you know yourself if yes then go ahead on day 4 else revise the inputs given as per assessment results.

Personality and Work Style

Day 4 : ____/_____/________ (DD / MM / YYYY) – 4 Hours

Rework on Resume and cover letter based on detailed explanation given in Chapter 2

Resume Revamp

1: Self-Assessment and Goal Setting

- **Task:** Evaluate your current resume and identify areas for improvement.
- **Goal:** Set clear objectives for the week, such as highlighting key skills, and achievements, and tailoring the resume for specific job roles.

2: Skill Identification and Keywords

- **Task:** Identify your core skills and competencies relevant to your target roles.
- **Goal:** Incorporate relevant keywords from job descriptions into your resume to enhance visibility to Applicant Tracking Systems (ATS).

3: Achievements and Impact Statements

- **Task:** Focus on quantifying your achievements in previous roles.
- **Goal:** Craft impactful statements that showcase the tangible results of your contributions.

4: Format and Design

- **Task:** Explore resume templates and choose a clean, professional design.
- **Goal:** Ensure your resume is visually appealing and easy to read.

5: Cover Letter Creation

- **Task:** Draft a cover letter template highlighting your motivation and key strengths.
- **Goal:** Have a flexible cover letter that can be tailored for different job applications.

6: Peer Review

- **Task:** Share your resume with a friend or mentor for feedback.

- **Goal:** Obtain constructive feedback to refine your resume further.

7: Final Review and Uploads

- **Task:** Conduct a final review, proofread, and make necessary adjustments.
- **Goal:** Upload your refined resume to job portals such as LinkedIn, Naukri, and Indeed.

Companies identification

Day 5 : _____/______/________ (DD / MM / YYYY) – 4 Hours

List down 30 companies as per your skills and personality traits

1. 11. 21.
2. 13. 22.
3. 14. 23.
4. 15. 24.
5. 15. 25.
6. 16. 26.
7. 17. 27.
8. 18. 28.
9. 19. 29.
10. 20. 30.

To optimize keywords at Google and receive notifications related to those keywords via Gmail, you can follow these steps:

Step 1: Create Google Alerts

1. Open Google Alerts: Go to the Google Alerts website (https://www.google.com/alerts) or search for "Google Alerts" on Google.

2. Enter Keywords: In the search bar, enter the keywords or phrases you want to receive notifications for. Be specific and use relevant terms related to your interests or industry.

3. Customize Settings: Click on "Show Options" to customize your alert settings. You can choose options such as:

- Sources: Select where Google should search for results (e.g., news, blogs, web, videos).

- Language: Choose the language of the content you want to receive alerts for.

- Region: Specify the region for which you want to receive alerts (optional).

- Frequency: Select how often you want to receive alerts (e.g., as-it-happens, once a day, once a week).

4. Preview Alerts: After customizing your settings, you can preview the alerts to see the type of results you'll receive.

5. Create Alert: Once satisfied with your settings, click on the "Create Alert" button.

Step 2: Manage Google Alerts in Gmail

1. Access Gmail: Open your Gmail account in a web browser.

2. Create Label for Alerts: To keep your alerts organized, create a label specifically for Google Alerts. You can do this by clicking on the "More" option in the left sidebar, selecting "Create new label," and entering a name for your label (e.g., Google Alerts).

3. Set up Filters (Optional): If you want to further organize your alerts or apply specific actions to them (e.g., marking them as important, or applying a specific label), you can create filters in Gmail. Go to Settings > See all settings > Filters and Blocked Addresses > Create a new filter.

4. Manage Alerts: You will start receiving alerts related to your keywords directly in your Gmail inbox. You can click on an alert to view the corresponding news article or web page.

Step 3: Review and Adjust Alerts as Needed

1. Review Alerts: Regularly review the alerts you receive in your Gmail inbox to stay updated on relevant news and information.

2. Adjust Settings: If you find that you're receiving too many alerts or not enough relevant results, you can adjust your alert settings accordingly. Go to the Google Alerts website to modify your existing alerts or create new ones with different keywords or settings.

3. Refine Keywords: Continuously refine your keyword selection to ensure you're receiving notifications for the most relevant content.

By following these steps, you can optimize keywords at Google and receive notifications via Gmail whenever news related to those keywords is published online. This can help you stay informed and up-to-date on topics of interest or importance to you.

Second assignment: Networking and follow these company

Networking with identified companies involves engaging with them through various online platforms to stay updated on their activities and job opportunities. Here's how you can do it:

Step 1: Visit Company Websites

1. Subscribe to Newsletter: Explore the company website and look for options to subscribe to their newsletter or mailing list. Typically, you'll find a subscription form on the homepage or in the footer section. Enter your email address and follow the prompts to subscribe.

Step 2: Follow on Social Media

1. LinkedIn:

- Search for the company's LinkedIn page using the search bar.

- Click on the company's official page and click the "Follow" button to stay updated on their posts and job openings.

- Engage with their content by liking, commenting, and sharing relevant posts to establish visibility and connections.

2. Twitter:

- Search for the company's official Twitter handle using the search bar.

- Follow the company's Twitter account to receive updates on news, events, and job opportunities.

- Engage with their tweets by liking, retweeting, and commenting to show interest and build connections.

Step 3: Job Portals

1. Create Accounts:

- Sign up for accounts on popular job portals such as LinkedIn, Naukri, Indeed, and Glassdoor if you haven't already.

- Complete your profile and upload your resume to make yourself visible to recruiters.

2. Set Job Alerts:

- Use the job search feature on these portals to find job openings at your target companies.

- Set up job alerts with specific keywords and preferences related to your desired roles and industries.

- Receive notifications via email or app alerts whenever new job openings matching your criteria are posted by the identified companies.

Step 4: Engage and Network

1. Interact with Content:

- Regularly visit the company's website and social media profiles to stay updated on the latest news, products, and initiatives.

- Engage with their content by liking, commenting, and sharing posts relevant to your interests and expertise.

2. Connect with Employees:

- Use LinkedIn to connect with employees of the target companies, especially those in your desired department or role.

- Send personalized connection requests mentioning your interest in their company and willingness to learn more about their experiences.

3. Attend Virtual Events:

- Keep an eye out for virtual events, webinars, or workshops hosted by the companies you're interested in.

- Register for these events to gain insights, network with professionals, and showcase your interest in the company.

By following these steps, you can effectively network with identified companies, stay informed about their activities, and position yourself for potential job opportunities within those organizations. Remember to maintain professionalism and genuine interest in your interactions to build meaningful connections.

Job Description Identification and Job Application

Day 6 : ____/_____/________ (DD / MM / YYYY) – 4 Hours

Company Name | Job Title | Job Opening (Posting Date) | Job Portal

1.
2.
3.
4.
5.
6.
7.
8.
9.
10.
11.
12.
13.
14.
15.
16.
17.
18.
19.
20.
21.

22.

23.

24.

25.

26.

27.

28.

29.

30.

Keeping track of job openings, job descriptions, important dates, and announcements is crucial for effective job application and staying informed about potential opportunities. Here's a structured approach to managing this process:

Step 1: Job Description Identification

1. Regularly Monitor Company Websites:

- Visit the careers or jobs section of the company websites you're interested in.

- Keep an eye out for new job postings and updates.

2. Set up Job Alerts:

- Utilize job portals like LinkedIn, Naukri, and Indeed to set up job alerts for specific companies and positions.

- Receive notifications whenever new job openings matching your criteria are posted.

3. Review Job Descriptions:

- Read job descriptions carefully to understand the required qualifications, skills, and responsibilities.

- Note down key requirements and responsibilities relevant to your expertise and experience.

Step 2: Job Application Preparation

1. Tailor Your Resume and Cover Letter:

- Update your resume and cover letter based on the latest job description and requirements.

- Highlight relevant skills, experiences, and achievements that align with the job requirements.

- Customize your application to emphasize your suitability for the specific role.

2. Keep Track of Changes:

- Maintain a record of changes in job descriptions and requirements over time.

- Note any subtle shifts in project requirements or core responsibilities.

Step 3: Tracking Important Dates and Announcements

1. Calendar Reminders:

- Use calendar apps or reminders to track important dates such as application deadlines, interview schedules, and result announcements.

- Set up notifications to ensure you don't miss any deadlines or updates.

2. Stay Informed:

- Regularly check company websites, social media channels, and job portals for updates on recruitment processes and announcements.

- Follow the companies on LinkedIn and Twitter to receive real-time updates and news.

3. Review Previous Year's Results:

- Analyze the results published from the previous year's recruitment process.

- Identify any patterns or trends in the hiring process and outcomes.

Step 4: Job Application and Follow-up

1. Submit Applications Promptly:

- Apply for job openings promptly to increase your chances of consideration.

- Ensure your application materials are tailored to the specific job requirements.

2. Follow-up:

- If you don't hear back within a reasonable time frame, consider following up with a polite email or phone call to inquire about the status of your application.

- Demonstrate your continued interest in the position and company.

By following this systematic approach, you can effectively track job openings, tailor your applications, and stay informed about important dates and announcements. This proactive approach increases your chances of success in the job application process.

We Kept this day for no work means enjoying or revisiting any pending day task.

Day 7 : ____/_____/_______ (DD / MM / YYYY) – 4 Hours

CHAPTER SEVEN

Week 2: Technical Triumph

Practice Common Technical Questions:

Prepare responses to common technical questions, including last year question papers and basic technical questions

Day 7 : ____/____/______ (DD / MM / YYYY) – 4 Hours

Assignment: Daily study sessions and revision lessons to reinforce technical knowledge, we advise revisit Chapter 3 if you failed to answer below questions

Mock Technical Question – 50 Marks 30 Minutes

Here are 50 multiple-choice questions related to the power sector

These questions cover various aspects of the power sector, including different types of power plants, renewable energy sources, electrical components, and environmental considerations.

1. Which of the following is the primary source of energy for thermal power plants?

a) Coal

b) Wind

c) Solar

d) Natural Gas

2. Which type of power plant produces electricity by harnessing the kinetic energy of flowing water?

a) Hydroelectric

b) Geothermal

c) Biomass

d) Nuclear

3. What is the typical voltage level of electricity generated by a power plant before transmission?

a) 110V

b) 230V

c) 11kV

d) 400kV

4. Which component of a nuclear power plant is responsible for initiating and controlling the nuclear reaction?

a) Reactor core

b) Turbine

c) Generator

d) Control rods

5. What is the process of converting sunlight into electricity called?

a) Photovoltaic

b) Thermoelectric

c) Hydropower

d) Wind energy

6. Which of the following is a renewable source of energy?

a) Coal

b) Natural gas

c) Solar

d) Oil

7. What is the unit of measurement for electrical power?

a) Watts

b) Joules

c) Volts

d) Amperes

8. Which type of power plant produces electricity by burning organic materials such as wood or agricultural waste?

a) Biomass

b) Nuclear

c) Geothermal

d) Wind

9. Which of the following is a byproduct of burning coal in a thermal power plant?

a) Carbon dioxide

b) Oxygen

c) Nitrogen

d) Hydrogen

10. What is the purpose of a transformer in an electrical power system?

a) To increase voltage

b) To decrease voltage

c) To convert DC to AC

d) To store electrical energy

11. Which of the following is NOT a component of a gas turbine power plant?

a) Combustion chamber

b) Turbine

c) Boiler

d) Compressor

12. What is the function of a capacitor in an electrical circuit?

a) To store electrical energy

b) To regulate voltage

c) To convert AC to DC

d) To amplify current

13. Which of the following renewable energy sources relies on the gravitational force of the Earth and Moon?

a) Tidal energy

b) Solar energy

c) Wind energy

d) Geothermal energy

14. Which material is commonly used as a coolant in nuclear power plants?

a) Water

b) Oil

c) Mercury

d) Sodium

15. What is the process of removing impurities from raw coal called?

a) Pulverization

b) Combustion

c) Gasification

d) Coal beneficiation

16. Which of the following is a disadvantage of using nuclear power as an energy source?

a) Greenhouse gas emissions

b) High initial capital costs

c) Limited availability of fuel

d) Environmental pollution

17. What is the primary function of a steam turbine in a thermal power plant?

a) To generate electricity

b) To regulate voltage

c) To produce steam

d) To compress air

18. Which of the following renewable energy sources is dependent on geographic location and weather patterns?

a) Solar energy

b) Biomass energy

c) Geothermal energy

d) Wind energy

19. What is the purpose of a step-up transformer in an electrical power system?

a) To increase voltage

b) To decrease voltage

c) To convert AC to DC

d) To store electrical energy

20. Which of the following materials is commonly used as a fuel in nuclear reactors?

a) Uranium

b) Coal

c) Natural gas

d) Diesel

21. What is the typical efficiency range of a modern gas turbine power plant?

a) 20-30%

b) 40-50%

c) 60-70%

d) 80-90%

22. Which of the following is a renewable source of energy derived from organic matter?

a) Biomass energy

b) Nuclear energy

c) Natural gas

d) Oil

23. What is the primary purpose of a boiler in a thermal power plant?

a) To generate steam

b) To generate electricity

c) To compress air

d) To cool the condenser

24. Which of the following is a characteristic of alternating current (AC)?

a) Direction of current changes periodically

b) Flow of current is unidirectional

c) Voltage remains constant

d) Requires a battery for operation

25. What is the main advantage of using renewable energy sources over fossil fuels?

a) Lower cost

b) Limited availability

c) Reduced environmental impact

d) Higher energy density

26. Which of the following is NOT a component of a photovoltaic solar power system?

a) Inverter

b) Solar panels

c) Turbine

d) Battery

27. What is the function of a capacitor in an electrical circuit?

a) To store electrical energy

b) To regulate voltage

c) To convert AC to DC

d) To amplify current

28. Which of the following is a characteristic of direct current (DC)?

a) Flow of current is unidirectional

b) Direction of current changes periodically

c) Voltage remains constant

d) Requires alternating current for operation

29. What is the purpose of a condenser in a thermal power plant?

a) To convert steam into water

b) To produce steam

c) To regulate voltage

d) To store electrical energy

30. Which of the following renewable energy sources is derived from the Earth's internal heat?

a) Solar energy

b) Wind energy

c) Geothermal energy

d) Tidal energy

31. What is the purpose of a control rod in a nuclear reactor?

a) To initiate the nuclear reaction

b) To regulate the flow of coolant

c) To absorb excess neutrons

d) To generate electricity

32. Which of the following is a byproduct of burning natural gas in a power plant?

a) Carbon dioxide

b) Oxygen

c) Nitrogen

d) Water vapor

33. What is the function of a generator in an electrical power system?

a) To convert mechanical energy into electrical energy

b) To regulate voltage

c) To store electrical energy

d) To amplify current

34. Which of the following is a disadvantage of using wind energy as an energy source?

a) Limited availability

b) High initial capital costs

c) Environmental pollution

d) Intermittency of wind

35. What is the process of converting biomass into biofuel called?

a) Combustion

b) Gasification

c) Fermentation

d) Distillation

36. Which of the following is a renewable energy source derived from the movement of ocean tides?

a) Solar energy

b) Wind energy

c) Tidal energy

d) Geothermal energy

37. What is the primary function of a transformer in an electrical power system?

a) To increase voltage

b) To decrease voltage

c) To convert DC to AC

d) To store electrical energy

38. Which of the following renewable energy sources relies on the Earth's natural heat?

a) Solar energy

b) Wind energy

c) Geothermal energy

d) Biomass energy

39. What is the main advantage of using hydroelectric power as an energy source?

a) Low environmental impact

b) High initial capital costs

c) Limited availability

d) Intermittency of water flow

40. Which of the following is a renewable energy source derived from organic waste materials?

a) Biomass energy

b) Solar energy

c) Geothermal energy

d) Nuclear energy

41. What is the primary function of a rectifier in an electrical circuit?

a) To convert AC to DC

b) To regulate voltage

c) To amplify current

d) To store electrical energy

42. Which of the following renewable energy sources relies on capturing heat from the Earth's interior?

a) Solar energy

b) Wind energy

c) Tidal energy

d) Geothermal energy

43. What is the purpose of a heat exchanger in a thermal power plant?

a) To transfer heat from steam to water

b) To regulate voltage

c) To store electrical energy

d) To amplify current

44. Which of the following materials is commonly used as a fuel in biomass power plants?

a) Wood

b) Coal

c) Uranium

d) Natural gas

45. What is the primary function of a turbine in an electrical power system?

a) To convert mechanical energy into electrical energy

b) To regulate voltage

c) To store electrical energy

d) To amplify current

46. Which of the following is a characteristic of renewable energy sources?

a) Limited availability

b) High environmental impact

c) Non-renewable

d) Sustainability

47. What is the purpose of a capacitor in an electrical circuit?

a) To store electrical energy

b) To regulate voltage

c) To convert AC to DC

d) To amplify current

48. Which of the following is a renewable energy source derived from the sun's rays?

a) Solar energy

b) Wind energy

c) Tidal energy

d) Geothermal energy

49. What is the function of a rectifier in an electrical circuit?

a) To convert AC to DC

b) To regulate voltage

c) To amplify current

d) To store electrical energy

50. Which of the following renewable energy sources relies on harnessing the energy of ocean waves?

a) Solar energy

b) Wind energy

c) Tidal energy

d) Geothermal energy

Answers : Here are the answers to the multiple-choice questions:

1. a) Coal

2. a) Hydroelectric

3. d) 400kV

4. a) Reactor core

5. a) Photovoltaic

6. a) Coal

7. a) Watts

8. a) Biomass

9. a) Carbon dioxide

10. a) To increase voltage

11. c) Boiler

12. a) To store electrical energy

13. a) Tidal energy

14. a) Water

15. d) Coal beneficiation

16. b) High initial capital costs

17. a) To generate electricity

18. d) Wind energy

19. a) To increase voltage

20. a) Uranium

21. b) 40-50%

22. a) Biomass energy

23. a) To generate steam

24. a) Direction of current changes periodically

25. c) Reduced environmental impact

26. c) Turbine

27. a) To store electrical energy

28. a) Flow of current is unidirectional

29. a) To convert steam into water

30. c) Geothermal energy

31. c) To absorb excess neutrons

32. a) Carbon dioxide
33. a) To convert mechanical energy into electrical energy
34. d) Intermittency of wind
35. b) Gasification
36. c) Tidal energy
37. a) To increase voltage
38. c) Geothermal energy
39. a) Low environmental impact
40. a) Biomass energy
41. a) To convert AC to DC
42. d) Geothermal energy
43. a) To transfer heat from steam to water
44. a) Wood
45. a) To convert mechanical energy into electrical energy
46. d) Sustainability
47. a) To store electrical energy
48. a) Solar energy
49. a) To convert AC to DC
50. c) Tidal energy

If you like similar way of testing your technical expertise then Here are some multiple-choice questions related to Electrical, Control & Instrumentation (C&I) Engineering, focusing on renewable energy:

1. Which of the following renewable energy sources relies on converting sunlight into electricity?

a) Wind energy
b) Hydroelectric energy
c) Geothermal energy
d) Solar energy

2. What is the primary function of an inverter in a solar photovoltaic system?

a) Converts DC power to AC power
b) Converts AC power to DC power
c) Regulates voltage levels
d) Stores electrical energy

3. Which of the following components is used to store excess electrical energy generated by renewable sources?

a) Capacitor
b) Transformer

c) Battery

d) Resistor

4. What type of energy storage technology is commonly used in grid-scale renewable energy projects?

a) Lead-acid batteries

b) Lithium-ion batteries

c) Pumped hydro storage

d) Flywheel energy storage

5. Which renewable energy technology involves harnessing the kinetic energy of moving air?

a) Solar energy

b) Biomass energy

c) Wind energy

d) Geothermal energy

6. What is the primary function of a supervisory control and data acquisition (SCADA) system in a renewable energy plant?

a) Regulates voltage levels

b) Monitors and controls plant operations

c) Converts AC power to DC power

d) Stores electrical energy

7. Which of the following renewable energy sources relies on capturing heat from the Earth's interior?

a) Solar energy

b) Wind energy

c) Tidal energy

d) Geothermal energy

8. What is the purpose of a pitch control system in a wind turbine?

a) Regulates the angle of the turbine blades

b) Controls the speed of the generator

c) Converts AC power to DC power

d) Stores electrical energy

9. Which of the following is a characteristic of a grid-tied solar power system?

a) Requires batteries for energy storage

b) Operates independently of the utility grid

c) Exports excess electricity to the grid

d) Generates power only during daylight hours

10. What is the primary function of a grid interconnection system in a renewable energy project?

a) Converts AC power to DC power

b) Controls the flow of electricity between the plant and the grid

c) Regulates voltage levels within the plant

d) Stores electrical energy

11. Which renewable energy technology involves converting organic waste into biogas for electricity generation?

a) Biomass energy

b) Solar energy

c) Wind energy

d) Geothermal energy

12. What is the primary function of a power converter in a wave energy conversion system?

a) Regulates voltage levels

b) Converts mechanical energy into electrical energy

c) Controls the flow of electricity to the grid

d) Stores electrical energy

13. Which of the following is a key component of a concentrated solar power (CSP) system?

a) Photovoltaic panels

b) Parabolic troughs

c) Wind turbines

d) Geothermal heat pumps

14. What is the primary function of a charge controller in a standalone solar power system?

a) Regulates voltage levels

b) Converts AC power to DC power

c) Prevents overcharging of batteries

d) Stores electrical energy

15. Which renewable energy technology involves using the temperature difference between hot and cold water to generate electricity?

a) Tidal energy

b) Hydroelectric energy

c) Geothermal energy

d) Biomass energy

Answers: Here are the answers to the multiple-choice questions:

1. d) Solar energy

2. a) Converts DC power to AC power

3. c) Battery

4. c) Pumped hydro storage

5. c) Wind energy

6. b) Monitors and controls plant operations

7. d) Geothermal energy

8. a) Regulates the angle of the turbine blades

9. c) Exports excess electricity to the grid

10. b) Controls the flow of electricity between the plant and the grid

11. a) Biomass energy

12. b) Converts mechanical energy into electrical energy

13. b) Parabolic troughs

14. c) Prevents overcharging of batteries

15. c) Geothermal energy

Revise what you know and practice the technical expertise on latest industry trends

Day 8 : ____/_____/_______ (DD / MM / YYYY) – 4 Hours

Day 9 : ____/_____/_______ (DD / MM / YYYY) – 4 Hours

Use online learning tools and sites

Here are 10 reliable websites where you can find resources for technical revision related to power plants:

1. Power Magazine (www.powermag.com): Provides news, articles, and technical information related to power generation, including coal, gas, nuclear, and renewable energy.

2. International Atomic Energy Agency (IAEA) (www.iaea.org): Offers publications, reports, and technical documents on nuclear power plant operation, safety, and regulation.

3. National Renewable Energy Laboratory (NREL) (www.nrel.gov): Offers research, data, and publications on renewable energy technologies, including solar, wind, and biomass.

4. International Energy Agency (IEA) (www.iea.org): Provides reports, statistics, and analysis on energy markets, policies, and technologies, including those related to power generation.

5. The Institution of Engineering and Technology (IET) (www.theiet.org): Offers technical articles, webinars, and publications covering various aspects of electrical engineering, including power generation and distribution.

6. American Society of Mechanical Engineers (ASME) (www.asme.org): Provides resources, standards, and publications related to mechanical engineering, including power plant design, operation, and maintenance.

7. IEEE Power & Energy Society (PES) (www.ieee-pes.org): Offers technical articles, conferences, and standards related to power and energy engineering, including power plant technologies and systems.

8. Energy.gov (www.energy.gov): Provides information, reports, and resources on energy-related topics, including power generation, efficiency, and sustainability.

9. Power-technology.com (www.power-technology.com): Offers news, articles, and industry insights on power generation technologies, including thermal, hydro, and renewable energy.

10. The World Nuclear Association (WNA) (www.world-nuclear.org): Provides information, reports, and resources on nuclear energy, including reactor technology, safety, and sustainability.

These websites offer a wealth of resources, including articles, publications, reports, and technical documents, to help you revise and stay updated on power plant technologies and practices.

Day 10 : ____/_____/_______ (DD / MM / YYYY) – 4 Hours

Assignment: Daily study sessions and revision lessons to reinforce technical knowledge, we advise revisit Chapter 3 if you failed to answer below questions

Mock Technical Question – 50 Marks 30 Minutes

50 multiple-choice questions based on Thermal Power Plant Engineering and Operation

1. What is the primary source of energy used in a thermal power plant?

a) Wind

b) Solar

c) Coal

d) Natural gas

2. What is the purpose of a boiler in a thermal power plant?

a) To generate electricity

b) To convert water into steam

c) To store fuel

d) To regulate temperature

3. Which component of a thermal power plant is responsible for converting mechanical energy into electrical energy?

a) Turbine

b) Boiler

c) Condenser

d) Generator

4. What is the function of a condenser in a thermal power plant?

a) To increase the pressure of steam

b) To decrease the temperature of steam

c) To convert steam into water

d) To store electrical energy

5. Which of the following is the primary fuel used in thermal power plants?

a) Uranium

b) Natural gas

c) Coal

d) Sunlight

6. What is the function of a turbine in a thermal power plant?

a) To convert mechanical energy into electrical energy

b) To regulate voltage

c) To store energy

d) To cool down steam

7. What is the role of a condensate pump in a thermal power plant?

a) To increase the pressure of steam

b) To remove condensate from the condenser

c) To regulate temperature

d) To control the flow of steam

8. Which component of a thermal power plant is responsible for removing impurities from feedwater?

a) Boiler

b) Turbine

c) Deaerator

d) Generator

9. What is the primary function of a superheater in a thermal power plant?

a) To increase the pressure of steam

b) To decrease the temperature of steam

c) To increase the temperature of steam

d) To store energy

10. Which type of turbine is commonly used in thermal power plants?

a) Pelton turbine

b) Francis turbine

c) Kaplan turbine

d) Steam turbine

11. What is the main purpose of the economizer in a thermal power plant?

a) To reduce boiler fuel consumption

b) To regulate steam temperature

c) To remove impurities from feedwater

d) To increase electrical efficiency

12. Which component of a thermal power plant is responsible for converting steam back into water?

a) Turbine

b) Condenser

c) Boiler

d) Generator

13. What is the function of a steam drum in a thermal power plant?

a) To regulate steam pressure

b) To store steam

c) To increase steam temperature

d) To cool down steam

14. Which of the following is NOT a type of boiler commonly used in thermal power plants?

a) Fire-tube boiler

b) Water-tube boiler

c) Electric boiler

d) Biomass boiler

15. What is the primary purpose of a reheater in a thermal power plant?

a) To increase steam temperature

b) To decrease steam pressure

c) To remove impurities from steam

d) To store energy

16. Which component of a thermal power plant is responsible for controlling the flow of steam to the turbine?

a) Condenser

b) Deaerator

c) Governor

d) Generator

17. What is the primary advantage of a combined cycle power plant?

a) Higher efficiency

b) Lower capital cost

c) Simplicity of operation

d) Lower environmental impact

18. Which type of turbine extracts energy from both high-pressure and low-pressure steam?

a) Impulse turbine

b) Reaction turbine

c) Extraction turbine

d) Non-condensing turbine

19. What is the primary function of a deaerator in a thermal power plant?

a) To regulate steam pressure

b) To remove oxygen and other gases from feedwater

c) To control steam temperature

d) To increase steam velocity

20. Which of the following is NOT a method of cooling used in thermal power plants?

a) Air cooling

b) Water cooling

c) Oil cooling

d) Gas cooling

21. What is the main advantage of a once-through boiler system?

a) Lower maintenance requirements

b) Higher efficiency

c) Greater fuel flexibility

d) Simplicity of operation

22. Which component of a thermal power plant is responsible for regulating steam pressure?

a) Boiler

b) Condenser

c) Governor

d) Turbine

23. What is the function of an air preheater in a thermal power plant?

a) To increase the temperature of combustion air

b) To decrease the temperature of combustion air

c) To remove impurities from combustion air

d) To regulate the flow of combustion air

24. Which of the following is NOT a type of steam turbine commonly used in thermal power plants?

a) Impulse turbine

b) Reaction turbine

c) Axial-flow turbine

d) Centrifugal turbine

25. What is the primary advantage of a fluidized bed combustion boiler?

a) Higher combustion efficiency

b) Lower fuel flexibility

c) Greater steam production

d) Smaller footprint

26. Which component of a thermal power plant is responsible for reducing the temperature of exhaust gases?

a) Boiler

b) Turbine

c) Condenser

d) Air preheater

27. What is the main purpose of a steam trap in a thermal power plant?

a) To remove condensate from the steam line

b) To regulate steam pressure

c) To increase steam velocity

d) To decrease steam temperature

28. Which of the following is NOT a method of steam generation used in thermal power plants?

a) Pulverized coal combustion

b) Fluidized bed combustion

c) Circulating fluidized bed combustion

d) Biomass gasification

29. What is the function of a bypass valve in a thermal power plant?

a) To regulate steam pressure

b) To control steam flow

c) To divert steam around the turbine

d) To increase steam temperature

30. What is the primary advantage of a natural circulation boiler?

a) Lower maintenance requirements

b) Greater fuel flexibility

c) Higher combustion efficiency

d) Simplicity of operation

31. Which component of a thermal power plant is responsible for removing ash and other solid residues from flue gas?

a) Economizer

b) Air preheater

c) Electrostatic precipitator

d) Flue gas desulfurization system

32. What is the primary function of a draft fan in a thermal power plant?

a) To regulate steam pressure

b) To control air flow

c) To increase steam temperature

d) To
decrease steam velocity

33. Which type of turbine extracts energy from steam at atmospheric pressure?

a) Impulse turbine

b) Reaction turbine

c) Extraction turbine

d) Non-condensing turbine

34. What is the main advantage of a nuclear power plant compared to a thermal power plant?

a) Lower capital cost

b) Higher efficiency

c) Lower environmental impact

d) Greater fuel flexibility

35. Which of the following is NOT a type of cooling tower commonly used in thermal power plants?

a) Natural draft cooling tower

b) Mechanical draft cooling tower

c) Induced draft cooling tower

d) Forced draft cooling tower

36. What is the primary function of a de-superheater in a thermal power plant?

a) To increase steam pressure

b) To decrease steam temperature

c) To remove impurities from steam

d) To regulate steam flow

37. Which component of a thermal power plant is responsible for controlling steam flow to the turbine?

a) Boiler

b) Turbine

c) Governor

d) Generator

38. What is the primary purpose of a regenerative feedwater heater in a thermal power plant?

a) To increase boiler efficiency

b) To decrease steam temperature

c) To remove impurities from feedwater

d) To regulate steam pressure

39. Which of the following is NOT a type of coal commonly used in thermal power plants?

a) Bituminous coal

b) Sub-bituminous coal

c) Lignite coal

d) Anthracite coal

40. What is the main advantage of a pulverized coal combustion boiler?

a) Higher combustion efficiency

b) Lower fuel flexibility

c) Greater steam production

d) Smaller footprint

41. Which component of a thermal power plant is responsible for removing moisture from feedwater?

a) Boiler

b) Turbine

c) Deaerator

d) Generator

42. What is the primary function of a re-circulation pump in a thermal power plant?

a) To increase steam pressure

b) To decrease steam temperature

c) To regulate steam flow

d) To remove impurities from steam

43. Which type of turbine is commonly used in combined cycle power plants?

a) Impulse turbine

b) Reaction turbine

c) Extraction turbine

d) Non-condensing turbine

44. What is the main advantage of a once-through cooling system compared to a recirculating cooling system?

a) Lower water consumption

b) Higher cooling efficiency

c) Greater environmental impact

d) Smaller footprint

45. Which component of a thermal power plant is responsible for regulating steam temperature?

a) Boiler

b) Turbine

c) Superheater

d) Generator

46. What is the primary function of a sootblower in a thermal power plant?

a) To remove impurities from feedwater

b) To remove ash from boiler tubes

c) To control steam flow

d) To regulate steam pressure

47. Which of the following is NOT a method of steam generation used in thermal power plants?

a) Fluidized bed combustion

b) Circulating fluidized bed combustion

c) Gasification

d) Biomass combustion

48. What is the function of an economizer in a thermal power plant?

a) To increase boiler efficiency

b) To decrease steam temperature

c) To remove impurities from feedwater

d) To regulate steam pressure

49. What is the primary advantage of a natural draft cooling tower?

a) Lower operating cost

b) Higher cooling efficiency

c) Greater environmental impact

d) Smaller footprint

50. Which component of a thermal power plant is responsible for controlling the flow of fuel to the boiler?

a) Turbine

b) Boiler

c) Fuel pump

d) Generator

Answers :

1. c) Coal

2. b) To convert water into steam

3. d) Generator

4. b) To decrease the temperature of steam

5. c) Coal

6. a) To convert mechanical energy into electrical energy

7. b) To remove condensate from the condenser

8. c) Deaerator

9. c) To increase the temperature of steam

10. d) Steam turbine

11. a) To reduce boiler fuel consumption

12. b) To convert water into steam

13. c) Higher combustion efficiency

14. c) Natural gas

15. a) To convert DC power to AC power

16. c) Geothermal energy

17. a) It is predictable and reliable

18. b) Concentrated solar power (CSP)

19. a) It produces no emissions

20. b) To remove ash from boiler tubes

21. a) To regulate steam pressure

22. a) Higher efficiency

23. c) Water cooling

24. b) Wind energy

25. a) Tidal energy

26. c) To store energy

27. b) It has low installation costs

28. a) Francis turbine

29. c) To divert steam around the turbine

30. d) Simplicity of operation

31. c) To remove impurities from feedwater

32. b) To control air flow

33. a) Impulse turbine

34. c) Lower environmental impact

35. d) Forced draft cooling tower

36. b) To decrease steam temperature

37. c) Governor

38. a) To increase boiler efficiency

39. d) Smaller footprint

40. a) Higher combustion efficiency

41. c) Deaerator

42. c) To regulate steam flow

43. a) Impulse turbine

44. a) Lower water consumption

45. c) Superheater

46. b) To remove ash from boiler tubes

47. c) Gasification

48. a) To increase boiler efficiency

49. a) Lower operating cost

50. c) Fuel pump

Revise what you know and practice the technical expertise on latest industry trends

Day 11 : ____/_____/_______ (DD / MM / YYYY) – **4 Hours** (Marks scored _____/ 50)

Day 12 : ____/_____/_______ (DD / MM / YYYY) – **4 Hours** (Marks scored _____/ 50)

Day 13: ____/_____/_______ (DD / MM / YYYY) – **4 Hours** (Marks scored _____/ 50)

Note : In this chapter we only guide you how to make sample action plan, you can design technical papers based on your expertise and selection.

Here are some technical library websites that offer free access to students in India:

1. National Digital Library of India (NDLI): Provides a wide range of digital educational resources including textbooks, lectures, articles, and videos across various disciplines.

2. Directory of Open Access Books (DOAB) India: Offers open access to academic books covering various subjects, including technical fields.

3. Indian Institute of Technology (IIT) Open Courseware: Provides access to lecture notes, videos, and other educational materials from various IITs in India.

4. Indian Institute of Science (IISc) E-prints Repository: Offers access to research articles, conference papers, and theses from the Indian Institute of

Science.

5. Indian Council of Agricultural Research (ICAR) Open Access Repository: Provides access to research articles, reports, and publications in the field of agriculture and related sciences.

6. Digital Library of India (DLI): Offers access to a collection of digitized books, manuscripts, and journals from various academic institutions in India.

7. Open Access Journals India (OAJI): Provides access to a selection of open access journals covering various disciplines including technical fields.

CHAPTER EIGHT

WEEK 3 : MASTERING GROUP DISCUSSIONS

Daily exercises to improve communication and teamwork skills

Day 15: ____/____/_______ (DD / MM / YYYY) – 4 Hours

Assignment: Find a friend who physically participate with you, if not having then use your mobile as friend who can either record your activities else you can connect with other family / friends through vide calls.

Let's adapt the activities and role-playing scenarios for individual candidates preparing for group discussions in job selection:

Activity 1. Storytelling Sessions: 30 Minutes each candidate

- Activity: Practice storytelling by recounting personal anecdotes or experiences related to professional achievements or challenges.

- Role-play: Enact scenarios where candidates simulate job-related situations, such as presenting a project proposal or handling a client meeting.

Note : Record the session then you can evaluate , based on your performance

Activity 2. Group Discussions: - 60 Minutes for Group

- Activity: Participate in mock group discussions on topics relevant to the job or industry, focusing on effective communication and collaborative problem-solving.

- Role-play: Take on different roles within the group, such as discussion leader, summarizer, or devil's advocate, to practice active participation and facilitation skills.

Note : Moderator will be judge the group and evaluate them , refer chapter 4 for more details on Group Discussion

Day 16: ____/____/_______ (DD / MM / YYYY) – 4 Hours

Activity 3. Team-building Games: 60 Minutes for Group

- Activity: Engage in team-building exercises that emphasize cooperation and coordination, such as solving puzzles or completing group challenges.

- Role-play: Assume leadership roles within the team, demonstrating the ability to motivate and inspire others while working towards a common goal.

Activity 4. Role-playing Exercises: 60 Minutes for Group

- Activity: Role-play job interview scenarios where candidates take turns playing the role of interviewer and interviewee, practicing effective communication and interpersonal skills.

- Role-play: Act out challenging workplace scenarios, such as resolving conflicts with colleagues or delivering difficult feedback to a team member, to enhance problem-solving and conflict resolution abilities.

Day 17: ____/____/______ (DD / MM / YYYY) – 4 Hours

Activity 5 . Collaborative Projects: 60 Minutes for Group

- Activity: Collaborate on projects or case studies with other candidates, focusing on division of tasks, communication of ideas, and integration of diverse perspectives.

- Role-play: Assign specific roles within the project team, such as project manager, researcher, or presenter, to develop leadership and teamwork skills in a simulated work environment.

Activity 6. Group Problem-solving Challenges: 60 Minutes for Group

- Activity: Participate in group problem-solving activities that require candidates to analyze issues, generate solutions, and reach consensus within a team setting.

- Role-play: Practice decision-making and negotiation skills by assuming different roles in problem-solving scenarios, such as mediator, advocate, or decision-maker.

Day 18: ____/____/______ (DD / MM / YYYY) – 4 Hours

Activity 7. Outdoor Activities:

- Activity: Engage in outdoor team-building activities, such as ropes courses or orienteering challenges, to foster trust, communication, and collaboration among candidates.

- Role-play: Lead outdoor expeditions or group challenges, demonstrating leadership and decision-making abilities in dynamic environments.

Day 19: ____/____/______ (DD / MM / YYYY) – 4 Hours

Activity 8. Team-based Story Writing:

- Activity: Collaborate on writing projects or case studies with other candidates, focusing on clear communication, logical structure, and effective teamwork.

- Role-play: Assume different roles within the writing team, such as editor, researcher, or content creator, to develop collaborative writing and editing skills.

Activity 9 : Reflective Discussions:

- Activity: Reflect on group interactions and experiences during debrief sessions, discussing strengths, areas for improvement, and strategies for future success.

- Role-play: Engage in reflective dialogues with peers, taking turns as both speaker and listener, to develop self-awareness, empathy, and effective communication skills in group settings.

Day 20: ____/_____/_______ (DD / MM / YYYY) – 4 Hours

Activity 10. Group Presentations:

- Activity: Prepare and deliver group presentations on topics relevant to the job or industry, focusing on effective communication, presentation skills, and teamwork.

- Role-play: Practice presenting and defending ideas in front of peers, taking on roles such as presenter, facilitator, or audience member, to enhance public speaking and persuasion abilities.

By adapting these activities and role-playing scenarios for individual candidates preparing for group discussions in job selection, individuals can enhance their communication, teamwork, and interpersonal skills, ultimately improving their chances of success in the recruitment process.

We Kept this day for no work means enjoy or revisit any pending day task.

Day 21 : ____/_____/_______ (DD / MM / YYYY) – 4 Hours

CHAPTER NINE

WEEK 4 – DEVELOPING INTERVIEW STRATEGIES

Techniques for managing interview anxiety and building confidence.

Getting over interview nervousness and boosting confidence are very important skills for people getting ready for job interviews. To help you do this, here are some helpful tips:

1. Preparation and Practice: Learn as much as you can about the company, the job, and the business to know what to expect.

—Rehearse how to answer common job questions by yourself and with a friend or mentor.

- Do practice interviews to get stronger in your confidence and experience with real jobs.

2. Positive Visualization: - Picture yourself doing well in the interview, answering questions with confidence and wowing the interviewer.

- Saying happy affirmations can help you believe in your skills and abilities even more.

3. Deep Breathing and Relaxation Techniques: — Do deep breathing exercises before and during the interview to calm down and feel less anxious.

– To calm your mind and body, try progressive muscle relaxation or mindfulness meditation.

4. Focus on the Present Moment: — During the interview, stay focused on the present moment by listening to the interviewer's questions and giving thoughtful answers.

—Do not dwell on mistakes made in the past or worry about what might happen in the future.

5. Power Posing: Show that you are sure of yourself by standing tall, shoulders back, and chin up. This will show the reporter and yourself that you are sure of yourself.

—Do power poses before the interview to feel more confident and less stressed.

6. Pre-Interview Rituals: - Find a way to feel centered and confident before the interview, like going for a brisk walk, listening to uplifting music, or practicing mindfulness.

7. Look at your strengths and accomplishments. This will help you remember the things you've done, the skills you have, and the qualifications you have that make you a good choice for the job.

- Think of examples of things you've done well in the past to show that you are qualified for the job.

8. Have realistic expectations. Know that it's normal to be nervous before an interview and that some worry can help you do your best.

- Be aware that you might not do a great job, but try your best and learn from the experience.

9. Get Help: — Before the interview, talk to friends, family, or teachers to get support and encouragement.

- If interview anxiety really gets in the way of your daily life, you might want to talk to a job counselor or therapist.

10. Post-Interview Reflection: - Think about how you did in the interview and what went well and what you could have done better.

- Use comments from practice interviews or debriefings to improve your interview skills for real interviews.

By using these techniques, job candidates can effectively deal with interview anxiety and boost their confidence, which will eventually improve their performance and raise their chances of getting the job.

Day 22 : ____/_____/________ (DD / MM / YYYY) – 4 Hours

Activity: Mock Interview with friend/family

Interview Duration: 60 Minutes

Mock interviews are a valuable and effective tool for preparing for actual job interviews. They provide candidates with a simulated interview experience, allowing them to practice their responses, refine their communication skills, and build confidence. Here's a step-by-step guide to conducting a mock interview activity:

Step 1: Set Up the Mock Interview Environment

1. Choose a Quiet Space:

- Select a quiet and distraction-free environment to mimic the setting of a real interview.

- Ensure good lighting and a neutral background for video mock interviews.

2. Dress Professionally:

- Encourage the candidate to dress as they would for an actual interview to create a realistic atmosphere.

Step 2: Define the Interviewer and Candidate Roles

1. Interviewer Role:

- Assign someone (a friend, colleague, or career advisor) to play the role of the interviewer.

- The interviewer should have a list of questions and scenarios prepared.

2. Candidate Role:

- The individual being interviewed should take on the role of the candidate.

- Provide them with a copy of the job description and background information about the company.

Step 3: Develop a List of Interview Questions

1. Customize Questions:

- Tailor the interview questions to match the specific job the candidate is preparing for.

- Include common behavioral, situational, and competency-based questions.

2. Prepare Follow-up Questions:

- Anticipate follow-up questions based on the candidate's responses to create a dynamic and realistic experience.

Step 4: Conduct the Mock Interview

1. Introduction:

- Begin the mock interview with a brief introduction, mirroring how an actual interview would start.

2. Ask Interview Questions:

- The interviewer should ask a series of questions related to the candidate's qualifications, experience, and skills.

- Encourage the candidate to respond as they would in a real interview.

3. Provide Feedback:

- After each question or scenario, the interviewer should provide constructive feedback on the candidate's responses.

- Highlight strengths and offer suggestions for improvement.

Step 5: Focus on Non-Verbal Communication

1. Body Language:

- Pay attention to the candidate's body language, including posture, eye contact, and gestures.

- Offer feedback on how non-verbal cues can be adjusted for a more confident presentation.

2. Voice Tone and Pace:

- Evaluate the candidate's tone of voice, pitch, and speaking pace.

- Suggest adjustments to convey professionalism and enthusiasm.

Step 6: Role Reversal

1. Switch Roles:

- Have the candidate and interviewer switch roles to give the candidate a chance to practice asking questions and evaluating responses.

2. Provide Guidance:

- Offer guidance on effective questioning techniques and how to probe for additional information.

Step 7: Debrief and Reflect

1. Debrief the Experience:

- After the mock interview, have a debriefing session where both the candidate and interviewer share their thoughts.

- Discuss specific areas of improvement and strengths.

2. Reflect on Feedback:

- Encourage the candidate to reflect on the feedback received and consider how they can apply it in future interviews.

Step 8: Repeat as Necessary

1. Repeat the Exercise:

- Conduct multiple mock interviews to address different types of questions and scenarios.

- Use different interviewers to simulate diverse interviewing styles.

2. Continuous Improvement:

- Encourage the candidate to continuously refine their responses and approach based on feedback received during mock interviews.

Benefits of Mock Interviews:

1. Skill Enhancement:

- Candidates can enhance their communication, problem-solving, and critical-thinking skills through regular practice.

2. Increased Confidence:

- Mock interviews help boost candidates' confidence by familiarizing them with the interview process.

3. Identification of Weaknesses:

- Candidates can identify areas of weakness and work on improving specific aspects of their interview performance.

4. Realistic Simulation:

- Simulating a real interview environment helps candidates better understand the dynamics of the process.

5. Feedback and Improvement:

- Constructive feedback from mock interviews allows candidates to make targeted improvements and adjustments.

By incorporating mock interviews into interview preparation, candidates can enhance their readiness, mitigate anxiety, and increase their chances of success in actual job interviews.

Day 23 : ____/_____/_______ (DD / MM / YYYY) – 4 Hours

- Shopping your selection of two sets of Interview Attire

Activity: Mock Interview with friend/family in Dress rehearsal with required documents
Interview Duration: 60 Minutes

- **Dressing professionally and projecting a positive demeanor.**

Making a good impression at a job interview is very important if you want to get the job. This image is formed by your credentials and experience as well as the way you act and look. This detailed guide will teach you how to dress properly and have a good attitude during job interviews. It will cover a lot of topics, including what to wear, how to look after yourself, how to communicate, and body language.

How to Dress Professionally: Making a Good First Impression

Professional dress is more than just wearing suits and ties. It shows that you are sure of yourself, your skills, and your appreciation for the chance. Here's how to look your best:

Find out about the company culture and understand it

Find out about the company's culture, principles, and dress code before you choose what to wear to the interview. You want to make sure that your outfit meets the needs of the business and shows that you are a professional.

Wear the Right Clothes

Choosing the right clothes means picking clothes that fit well, are clean, and are appropriate for the job and business you're applying for. Here is a list for both guys and women:

What Men Should Wear:

- Suit: Choose a well-fitted suit in a dark color like black, navy blue, or charcoal gray. Make sure the suit is the right size, neither too big nor too small.

This is the shirt: Put on a clean, long-sleeved dress shirt with a solid color or a light design. Shirts in white or light blue are always a good pick.

Tie: Pick a tie that goes with the suit and shirt and isn't too flashy. Pick designs that aren't too bright or busy instead of ones that are.

- Shoes: Wear black or brown dress shoes that are shiny and clean. Make certain they are clean and in good shape.

- Additional Items: Keep your accessories simple and stylish. A watch, simple cufflinks, and a briefcase or folder made for work can make you look more put-together.

What women wear:

- Suit: If you want to look sharp, wear a suit with a neutral-colored skirt or pants. Make sure the skirt is the right length, falling at or below the knee.

- This is a blouse: Wear a business blouse or button-down shirt in a solid color or a light pattern with the suit. Avoid tops with low cuts or styles that show too much skin.

- For shoes, choose grey or black boots or flats with closed toes. Avoid shoes with too many spikes or heels.

- Additional Items: Don't wear too much makeup or other accessories. Your outfit can look better with a watch, simple earrings, and a business-like purse or suitcase.

Grooming and taking care of yourself

Along with what you wear, how you look and how clean you are are also very important for making a good professional image. Here are some ideas:

- Hair Make sure your hair is clean, well-kept, and shaped nicely. If you want to look good, don't wear your hair in crazy colors or styles.

- Nails: Cut and clean your nails. If you want to wear nail polish, choose patterns that are neutral or simple.

- Hygiene for Men Take care of your own cleanliness by doing things like brushing your teeth, using deodorant, and staying away from strong perfumes and colognes.

• Facial Hair: If you have facial hair, make sure it is well-groomed and clipped. To look more put-together, you might want to shave or trim any facial hair that is too long.

Projecting a Positive Attitude: Being Confident Is Important

In addition to how you look, how you act is also very important during a job interview. How to act in a way that shows confidence, passion, and professionalism:

Good Body Language

More often than not, body language says more than words alone. If you want to show confidence through your body language, here are some tips:

- Make strong eye contact with the reporter to show that you are interested and confident. Don't stare, but make sure you look the person in the eye often during the interview.

Position: Keep your back straight and don't slouch or lean back in your chair. To look sure of yourself and confident, keep your shoulders back and your chest open.

- Gestures: To make your points clear and show passion, use open and natural gestures. Do not fidget or move your hands around too much, as this can be annoying.

- Smile: During the interview, smile kindly and honestly to show that you are friendly and easy to talk to. A smile can make you and the employer feel better.

Trust in Talking to Others

To make a good impression at a job interview, you need to be able to talk to people with confidence. Here's how to confidently and successfully talk to people:

- Be Clear and Sure of Yourself: Say what you want to say clearly and with confidence. Do not mumble or talk too fast, as this can make you look nervous or unsure of yourself.

- Use Positive Language: To show confidence and excitement, use positive and bold language. Don't use negative language or make jokes about yourself because they can hurt your trustworthiness.

- Active Listening: Nod, keep eye contact, and ask intelligent questions to show that you are actively listening. This shows that you're interested in what the person is saying and want to learn more.

- Answer with Full Trust: When you answer, be sure of yourself and say what you need to say quickly. Give specific cases and stories to back up your claims and show that you are qualified.

How to Act Professionally

Professional etiquette includes many different types of behavior and conduct that help you make a good impact. Here are some ways to act professionally during the interview:

- Be on Time: Being on time is very important for making a good first impression. Make sure you get to the interview spot early in case there are any delays.

- Dress Appropriately: Wear business-appropriate clothes that show you understand the company's standards and culture. Make sure your clothes are clean, fit well, and don't have any spots or wrinkles.

Please be polite and respectful: Be nice and respectful to everyone you meet, from the front desk worker to the interviewer.

When you treat others with respect and politeness, it looks good on your record.

- Express Your Thanks: At the end of the interview, thank the person who interviewed you for the chance to talk about the job. No matter what happens, thank them for their time and thought.

Final Thoughts

Dressing professionally and having a good attitude are important things to do to make a good impression at a job interview. You can improve your chances of success and make a lasting impact on potential employers by carefully choosing the right clothes, meticulously grooming yourself, and acting in a way that shows confidence and enthusiasm. Do some study on the company culture, use good body language, and speak with confidence to show that you are professional and qualified for the job. You can feel strong during job interviews and get the opportunities you want if you prepare, practice, and have the right attitude.

Day 24 : _____/______/________ (DD / MM / YYYY) – 4 Hours

Activity: Preparedness of documents to carry in India. When preparing for a job interview in India, it's essential to have a well-organized set of documents to present to the prospective employer.

Here's a comprehensive checklist to ensure you have all the necessary documents for a successful interview:

1. Resume:

- Multiple Copies:

- Print multiple copies of your updated resume.

- Ensure the information is accurate, and it aligns with the job description.

2. Cover Letter:
- Customized Letter:
- Have a tailored cover letter for the specific job you're applying for.
- Address it to the hiring manager or relevant person if possible.
3. Photo ID:
- Government-issued ID:
- Carry a valid government-issued photo ID, such as Aadhar Card, Passport, Driving License, or Voter ID.
- Ensure that the name on your ID matches the name on your resume.
4. Educational Documents:
- Academic Certificates:
- Carry your original academic certificates and degrees.
- Include your 10th, 12th, and any higher education certificates.
- Transcripts:
- If applicable, include transcripts or mark sheets from your educational institutions.
5. Work Experience Documents:
- Experience Certificates:
- Bring copies of experience certificates from previous employers.
- Include letters of recommendation if available.
- Offer Letters:
- Carry copies of offer letters from your current and previous jobs.
- Salary Slips:
- Include recent salary slips or payslips for verification.
6. References:
- Reference List:
- Prepare a list of professional references with contact information.
- Inform your references in advance.
7. Portfolio:
- Work Samples:
- If applicable, bring a portfolio showcasing relevant work samples.
- Include projects, presentations, or any tangible evidence of your skills.
8. Certifications and Licenses:
- Professional Certifications:
- Carry copies of relevant professional certifications.
- Include licenses required for the job, if any.
9. Passport-sized Photographs:
- Recent Photographs:

- Bring a few recent passport-sized photographs.
- They may be required for application forms or company records.

10. Online Presence:
- LinkedIn Profile:
- If you have a LinkedIn profile, ensure it is updated.
- Include the link on your resume.
- Online Portfolio:
- If applicable, provide a link to your online portfolio showcasing your work.

11. Other Documents:
- Offer Acceptance Letter:
- If applicable, carry a copy of your offer acceptance letter from the current employer.
- Resignation Letter:
- If you're currently employed, have a copy of your resignation letter.
- Address Proof:
- Carry a recent utility bill or rental agreement as proof of address.

12. Additional Items:
- Pen and Notepad:
- Carry a pen and notepad for taking notes during the interview.
- Questions for the Interviewer:
- Prepare a list of questions to ask the interviewer about the company, role, and team.

Tips:

1. Organization:
- Use a professional folder or portfolio to keep your documents organized and prevent any damage.

2. Check Company Requirements:
- Review the interview invitation or communication to see if there are any specific documents the company has requested.

3. Electronic Copies:
- Consider having electronic copies of your documents stored on a USB drive or in cloud storage for easy access.

4. Dress Code:
- Dress professionally as per the company's dress code and industry standards.

5. Arrive Early:

- Plan to arrive at least 15-30 minutes before the scheduled interview time.

By having a well-prepared set of documents, you not only present yourself as an organized and detail-oriented candidate but also ensure that you can respond promptly to any requests from the interviewer. This checklist covers the essentials but always check for any specific requirements from the company before the interview.

Video Record of Interview Pitch

Day 26 : ____/_____/_______ (DD / MM / YYYY) – 4 Hours

Activity: We need to record 2 minute video message which will help as covering letter and virtual resume

Creating a video introduction for your cover letter and resume can be an effective way to make a memorable impression on potential employers. Here are the steps to create a compelling video introduction:

1. Plan Your Content:

- Outline Your Message:

- Determine the key points you want to convey in your video introduction.

- Highlight your relevant skills, experiences, and accomplishments.

- Personalize Your Approach:

- Tailor your message to the specific job and company you're applying to.

- Address the hiring manager by name if possible.

2. Choose a Suitable Setting:

- Select a Professional Background:

- Choose a clean and clutter-free background for your video.

- Avoid distractions and ensure good lighting.

- Consider Your Appearance:

- Dress professionally and appropriately for the job you're applying for.

- Pay attention to grooming and personal presentation.

3. Set Up Your Recording Equipment:

- Use a Quality Camera:

- Use a high-definition camera for clear video quality.

- If using a smartphone, ensure it's stabilized and positioned correctly.

- Ensure Good Audio Quality:

- Use an external microphone or ensure your device's built-in microphone captures clear audio.

- Minimize background noise for optimal sound quality.

4. Write Your Script:

- Craft a Compelling Introduction:
- Start with a brief introduction of yourself and your background.
- Mention the position you're applying for and your interest in the company.
- Highlight Your Skills and Experience:
- Showcase your relevant skills, experiences, and achievements.
- Provide specific examples to illustrate your qualifications.
- Express Your Enthusiasm:
- Convey your passion for the role and company.
- Explain why you're excited about the opportunity and how you can contribute.
5. Record Your Video:
- Practice Before Recording:
- Rehearse your script several times to ensure fluency and confidence.
- Familiarize yourself with the camera and recording process.
- Record Multiple Takes:
- Record several takes of your video introduction to capture the best version.
- Take breaks between takes to review and make adjustments as needed.
6. Edit Your Video:
- Trim and Edit Footage:
- Use video editing software to trim any unnecessary footage and improve pacing.
- Add transitions, text overlays, or visual effects to enhance the presentation.
- Incorporate Visuals:
- Include relevant visuals such as screenshots of your work, graphs, or images that support your message.
- Add Personal Touches:
- Incorporate your personality and style into the video to make it engaging and authentic.
- Avoid over-editing and maintain a natural and professional tone.
7. Review and Finalize:
- Review Your Video:
- Watch the final version of your video introduction to ensure accuracy, clarity, and professionalism.
- Check for any technical issues or errors that need correction.
- Seek Feedback:

- Share your video with trusted friends, family members, or mentors for feedback.

- Consider their suggestions for improvement before finalizing the video.

8. Upload and Share:

- Choose a Platform:

- Upload your video introduction to a professional platform such as LinkedIn, Vimeo, or YouTube.

- Ensure the video is set to private or unlisted if you prefer to share it selectively.

- Include in Your Application:

- Embed the video link in your email cover letter or resume.

- Mention the video introduction in your application materials and provide instructions for viewing.

Creating a video introduction can set you apart from other candidates and give hiring managers a glimpse of your personality and communication skills. By following these steps and crafting a compelling video, you can make a memorable impression and increase your chances of landing the job opportunity.

Day 28,29,30: Mock Interviews

- **Assignment:** Practice common interview questions.
- **Milestone:** Conduct at least two mock interviews with feedback.

Most Important is Motivation and Encouragement:

- **Daily Affirmations:** Start each day with positive affirmations related to your career goals.
- **Weekly Reflections:** Reflect on your achievements and setbacks every Sunday. Celebrate victories and adjust your plan if needed.
- **Accountability Partner:** Share your action plan with a friend or mentor for ongoing support and encouragement

One day, there was a boy called Aarav who really wanted to have a certain job when he grew up. Aarav loved to draw and wanted to be a great artist, but no matter how many times he entered a drawing contest, he never won. He was sad about this and wanted to give up on his dream.

His grandmother told him a story about a small seed one day. "Aarav, picture yourself as a seed that wants to grow into a lovely flower," she told him. To grow strong, the seed needs water, sun, and time. To get better at what you love, you need practice, patience, and courage.

Aarav gave it some thought and chose not to give up. He started drawing every day, asking his friends for tips, and learning new ways to do things. He kept going even though some events turned him down.

Aarav chose to enter a big drawing event that was going on in town one day. He drew a picture that showed how much he cared and worked. The judges were amazed at how much Aarav had improved and how hard he had worked. Aarav won the event and became the most popular young artist in town.

This story about Aarav shows us that we should never give up, even when things don't go the way we want them to. With hard work and persistence, we can reach our goals, just like a tiny seed grows into a pretty flower over time. When you're having a hard time, think about Aarav's story and have faith in yourself.

CHAPTER TEN

Conclusion: Empowered for Success

Secrets to Interviewing Confidently and Standing Out as a Strong Candidate

Interviewing confidently and standing out as a strong candidate requires preparation, self-awareness, and effective communication. Here are some secrets to help you ace your interviews:

1. Research Thoroughly: Before the interview, research the company, its culture, values, and recent news or developments. Understanding the organization will help you tailor your responses and demonstrate your interest and enthusiasm.

2. Know Your Strengths: Reflect on your skills, experiences, and accomplishments. Identify your strengths and examples of how you've demonstrated them in previous roles. Be ready to discuss specific achievements that showcase your capabilities.

3. Practice Your Responses: Prepare answers to common interview questions, such as "Tell me about yourself" and "Why do you want to work here?" Practice articulating your responses clearly and confidently. Consider using the STAR method (Situation, Task, Action, Result) to structure your answers.

4. Dress Appropriately: Choose professional attire that is suitable for the company culture and industry. Dressing appropriately shows respect for the interviewer and demonstrates your professionalism.

5. Body Language Matters: Pay attention to your body language during the interview. Maintain good posture, make eye contact, and offer a firm handshake. Smile and nod to show engagement and interest in the conversation.

6. Be Authentic: Be yourself during the interview. Authenticity builds rapport and trust with the interviewer. Share genuine stories and experiences that highlight your personality and values.

7. Ask Thoughtful Questions: Prepare a list of questions to ask the interviewer. Thoughtful questions demonstrate your interest in the role and company while also providing valuable insights into the organization. Avoid asking questions that can be easily answered through research.

8. Show Enthusiasm: Express your enthusiasm for the opportunity and the company. Let the interviewer know why you're excited about the role and how you can contribute to the organization's success.

9. Follow-up: Send a thank-you email or note to the interviewer after the interview. Express gratitude for the opportunity and reiterate your interest in the position. This simple gesture can leave a positive impression and reinforce your enthusiasm for the role.

Comfortable Atmosphere for Both the Interviewer and Interviewee
Creating a comfortable atmosphere during an interview is essential for both the interviewer and the interviewee. Here's how to foster a relaxed and productive environment:

1. Start with a Warm Welcome: Greet the interviewees warmly and put them at ease from the moment they arrive. Offer them a drink and ensure they have a comfortable place to sit.

2. Set the Tone: Begin the interview with some friendly small talk to break the ice and establish rapport. This can help both parties feel more relaxed and open during the conversation.

3. Be Respectful and Professional: Treat the interviewee with respect and professionalism throughout the interview. Listen attentively, avoid interrupting, and maintain a positive and supportive demeanor.

4. Provide Clear Instructions: Clearly explain the format and structure of the interview process to the interviewee. Let them know what to expect and answer any questions they may have.

5. Encourage Open Communication: Create a safe space for the interviewee to share their thoughts, experiences, and questions. Encourage them to speak freely and express themselves openly.

6. Be Mindful of Body Language: Pay attention to your body language and nonverbal cues. Maintain open and welcoming body language to make the interviewee feel comfortable and valued.

7. Address Concerns Promptly: If the interviewee seems nervous or uncomfortable, address their concerns promptly and reassure them as needed. Offer encouragement and support throughout the interview process.

8. End on a Positive Note: Conclude the interview on a positive and encouraging note. Thank the interviewee for their time and interest in the position, and let them know what the next steps will be.

By creating a comfortable atmosphere during the interview, both the interviewer and interviewee can engage more effectively and have a more positive experience overall. This can lead to better communication, deeper insights, and ultimately, better hiring decisions.

Entering the job market with confidence is a crucial step toward achieving your career goals. Here's how you can build and maintain confidence as you navigate the job market:

1. Know Your Worth: Understand your skills, qualifications, and strengths. Recognize the value you bring to potential employers and believe in your abilities.

2. Set Clear Goals: Define your career objectives and set realistic goals for yourself. Having a clear direction will give you purpose and motivation as you enter the job market.

3. Prepare Thoroughly: Research companies, industries, and job roles that interest you. Tailor your resume and cover letter to highlight your relevant experiences and accomplishments.

4. Practice Interviewing: Practice answering common interview questions and participating in mock interviews to hone your skills. The more prepared you are, the more confident you'll feel during actual interviews.

5. Network Effectively: Build relationships with professionals in your desired industry through networking events, informational interviews, and online platforms like LinkedIn. Networking can open doors to new opportunities and boost your confidence.

6. Stay Positive: Focus on your strengths and achievements rather than dwelling on past rejections or setbacks. Maintain a positive attitude and approach each job application with optimism.

7. Seek Feedback: Be open to constructive feedback from mentors, peers, and interviewers. Use feedback as an opportunity for growth and improvement.

8. Practice Self-Care: Take care of your physical and mental well-being. Get enough rest, exercise regularly, and engage in activities that help you relax and recharge.

The Art of Asking and Listening is essential in the job market, both as a candidate and as an employee. Here's how you can master this skill:

1. Ask Thoughtful Questions: During interviews and networking conversations, ask insightful questions that demonstrate your interest and curiosity. Show that you've done your research and are genuinely interested in learning more.

2. Listen Actively: Pay close attention to what others are saying without interrupting. Practice active listening by nodding, maintaining eye contact, and asking follow-up questions to show your engagement.

3. Seek Clarification: If you're unsure about something, don't hesitate to ask for clarification. It's better to ask for clarification than to make assumptions that could lead to misunderstandings.

4. Empathize: Put yourself in the other person's shoes and try to understand their perspective. Show empathy and compassion in your interactions, and be receptive to others' viewpoints.

5. Take Notes: When appropriate, take notes during meetings and interviews to capture important information and demonstrate your attentiveness.

6. Reflect Before Responding: Before responding to a question or comment, take a moment to reflect on your thoughts. This will help you provide thoughtful and articulate responses.

7. Follow-Up: After conversations or meetings, follow up with a thank-you note or email. Express your appreciation for the opportunity to connect and reiterate key points from the discussion.

By mastering the art of asking thoughtful questions and listening actively, you'll not only build stronger relationships but also gain valuable insights that can help you succeed in the job market.

Follow-up and Post-Interview Etiquette are crucial aspects of the job search process. Here's how to navigate them effectively:

1. Send a Thank-You Note: After an interview or networking event, send a personalized thank-you note to express your gratitude for the opportunity to connect. Be sure to mention specific points from the conversation to

demonstrate your attentiveness.

2. Be Patient: Wait for a reasonable amount of time before following up on your application or interview. Avoid appearing too eager or impatient, but also don't hesitate to reach out if you haven't heard back within the expected timeframe.

3. Follow Up Strategically: When following up on your application or interview, be concise and respectful. Politely inquire about the status of your application or express your continued interest in the role.

4. Stay Professional: Maintain a professional tone in all your communications, whether it's through email, phone calls, or in-person interactions. Avoid being too informal or casual, especially when following up on job opportunities.

5. Keep Networking: Even after interviews or job applications, continue to nurture your professional relationships through networking. Stay in touch with contacts and keep them updated on your job search progress.

6. Handle Rejection Gracefully: If you receive a rejection letter or email, respond gracefully and thank the employer for considering your application. Stay positive and use the experience as an opportunity for growth and learning.

Salary Negotiation and Offer Letter negotiation can be intimidating, but with the right approach, you can negotiate effectively and secure a competitive salary. Here are some tips:

1. Research Market Rates: Research salary ranges for similar roles in your industry and location. Use online resources, salary surveys, and networking contacts to gather information about typical compensation packages.

2. Know Your Value: Understand your worth based on your skills, experience, and qualifications. Be prepared to articulate why you deserve a higher salary, citing specific examples of your contributions and achievements.

3. Consider Total Compensation: When evaluating a job offer, consider not only the base salary but also other benefits such as bonuses, healthcare, retirement plans, and vacation time. Negotiate for a comprehensive compensation package that meets your needs and priorities.

4. Practice Effective Communication: Approach salary negotiations with confidence and professionalism. Communicate your salary expectations and be prepared to negotiate respectfully and collaboratively.

5. Be Flexible: While it's important to advocate for fair compensation, be open to compromise and flexibility during negotiations. Consider factors

such as career growth opportunities, job responsibilities, and company culture in addition to salary.

6. Get it in Writing: Once you've reached an agreement on salary and benefits, request a formal offer letter outlining the terms and conditions of your employment. Review the offer carefully and seek clarification on any unclear points.

7. Seek Advice if Needed: If you're unsure about how to negotiate salary or evaluate a job offer, seek advice from mentors, career coaches, or HR professionals. They can provide valuable insights and guidance to help you navigate the negotiation process.

By mastering the art of negotiation and effectively communicating your value, you can secure a salary and benefits package that reflects your skills, experience, and contributions.

In conclusion, entering the job market with confidence, mastering the art of asking and listening, following up effectively, and negotiating salary and offer letters are essential skills for success. By cultivating these skills and approaches, you can navigate the job market with confidence and achieve your career goals.

Key Themes And Strategies Covered In The Book Include:

"Career Compounding : A 30-Days Blueprint for Landing Your Dream Job in the Power Sector" is a comprehensive guide designed to help individuals navigate the job market and secure their ideal position within the power sector. The book provides a structured, step-by-step approach that readers can follow over 30 days to achieve their career goals.

1. Self-Assessment: The book encourages readers to conduct a thorough self-assessment to identify their skills, strengths, and career aspirations. By understanding their unique abilities and interests, readers can align their job search efforts with positions that match their goals.

2. Market Research: Readers learn how to conduct research on the power sector job market, including identifying key players, trends, and job opportunities. By staying informed about industry developments, readers can position themselves as competitive candidates in the job market.

3. Resume and Cover Letter Optimization: The book offers guidance on crafting effective resumes and cover letters tailored to the power sector. Readers learn how to highlight their relevant skills, experiences, and achievements to capture the attention of potential employers.

4. Networking Strategies: Networking is emphasized as a critical component of the job search process. Readers learn how to build and leverage professional networks within the power sector to uncover hidden job opportunities and make meaningful connections.

5. Interview Preparation: The book provides tips and techniques for preparing for job interviews, including how to research companies, practice common interview questions, and present oneself confidently. Readers also learn strategies for effectively communicating their qualifications and value proposition during interviews.

6. Negotiation and Offer Acceptance: The book offers guidance on negotiating job offers and evaluating compensation packages within the power sector. Readers learn how to advocate for themselves professionally and secure favorable terms that align with their career objectives.

7. Continuous Learning and Growth: Throughout the book, the importance of continuous learning and professional development is emphasized. Readers are encouraged to stay informed about industry trends, acquire new skills, and pursue opportunities for advancement within the power sector.

Overall, "Career Compounding " provides readers with a comprehensive roadmap for navigating the job market and landing their dream job in the power sector. By following the actionable strategies outlined in the book, readers can maximize their chances of success and accelerate their career growth in this dynamic industry.